Patrick Chamberlain '86

At the crossroads
of cultures

In this series:

Cultures and time
Time and the philosophies

In preparation:

Time and the sciences

TIME and the PHILOSOPHIES

H. Aguessy M. Ashish Y. F. Askin Boubou Hama
H.-G. Gadamer L. Gardet A. Hasnaoui T. Honderich
A. Jeannière S. Karsz S. Ohe P. Ricœur
A. J. Toynbee J. Witt-Hansen

Published in 1977 by the United Nations
Educational, Scientific and Cultural Organization
7 Place de Fontenoy, 75700 Paris
Printed in Great Britain at The Benham Press, by
William Clowes & Sons Limited, London Colchester and Beccles

ISBN 92-3-101396-3
French edition: 92-3-201396-7

Printed in the United Kingdom

PREFACE

This book can be read independently of the preceding volume, *Cultures and Time*, and of the one to follow, *Time and the Sciences*.

Indeed, *Time and the Philosophies* must take up in turn some of the cultural aspects which were dealt with in the preceding volume.

Being itself a decanting of what is original and fundamental in cultures, philosophical analysis would ignore that basis only at the risk of slipping into the purely abstract approach of histories of philosophy, or into the anonymity of the great categories of thought. But Unesco's main concern in this project, known under the general title *At the Crossroads of Cultures*, is to feed philosophical analysis on the living substance of daily life and give it the savour of cultural diversity.

For it is out of the substance of everyday life and their own experimental fringe that the dialogue between the philosophies of time arises. Meeting 'at the crossroads', the complexity, singularity or even, sometimes, incommunicability of the cultures involved, make dialogue proportionately more arduous.

As the criss-cross of reciprocal influences becomes ever more dense and pervasive, each experience of time as lived inevitably conflicts with the consciousness others have of it, but each should also help to throw light on the ethical issues involved: hence the sections here on the pathology of time and its moral valuation.

The final section of the book offers a kind of portrait gallery of the soothsayer, the prophet, the guru, the modern leader, and the futurologist in their attitudes towards the philosophic problem of time—a dramatization, one might almost say, of the analyses of time, pushed to the extreme, with which the book begins.

NOTES ON THE AUTHORS

HONORAT AGUESSY. Born 1934, Cotonou, Benin. Ph.D. in Sociology, Letters and Humanities. *Chargé de Recherche*, Centre National de la Recherche Scientifique, France. Dean, Faculty of Letters, Arts and Humanities, National University of Benin.

Publications include: *Du Mode d'Existence de l'État sous Ghezo (1818–1859): Articulation de l'Instance Politique et de l'Instance Religieuse* (doctoral thesis, in press). Articles on 'La Religion Africaine comme Effet et Source de la Civilisation de l'Oralite', in *Présence Africaine*; 'Tradition Orale et Structures de Pensée: Essai de Méthodologie', in *Cahiers d'Histoire Mondiale*; 'Comment se dit la Parole (Valeur Propre de l'Oralité)'; 'La Divinité Legba et la Dynamique du Panthéon Vodû au Danhomê', in *Cahiers des Religions Africaines*; 'La Phase de la Négritude', in *Présence Africaine*; 'Tolérance dans les Religions Africaines', in *Axes*; 'Le Danhomê du 19e Siècle était-il une Société Esclavagiste?', in *Revue Francaise d'Études Politiques*; 'Histoire de la Pensée Africaine; "A Propos d'un Essai sur la Problématique de la Philosophie Africaine"', in *Axes*; 'Visions et Perceptions Traditionnelles', in *Introduction aux Cultures Africaines* (in preparation). In preparation: 'Étude de l'Histoire de l'Origine Culturelle Commune Ewe-Aja: Ife et Ketou Eu Égard à l'Histoire des Dassa, Save, Nago, Noatcha'.

SRI MADHAVA ASHISH. Born in 1920. Joined Uttar Brindaban Ashram (Mirtola, Almora District, U.P.), India, as disciple of the late Sri Krishna Prem in 1946, and took holy orders as a *Vairagi* (monk) of a Vishnava sect of Hinduism.

Publications include: co-author with Sri Krishna Prem of *Man, the Measure of all Things; Man, Son of Man*. Articles include: 'The Secret Doctrine as Contribution to World Thought' and 'Bondage as Freedom', in *The American Theosophist*; 'Sri Krishna Prem Through the Eyes of a Disciple', Foreword to *Initiation into Yoga. An Introduction to the Spiritual Life*, by Sri Krishna Prem.

YAKOV F. ASKIN. Professor, Head of the Department of Philosophy, University of Saratov, member of the Board of the Philosophical Society of the U.S.S.R.

Publications include: *The Problem of Time* (in Russian, Moscow, 1966; in Spanish, Montevideo, 1968; in Portuguese, Rio de Janeiro, 1969); *Philosophical*

Determinism and Scientific Knowledge (in Russian, Moscow, 1977); contributions to *Infinity and Universe* (in Russian, Moscow 1969; in Hungarian, Budapest, 1974); *Space, Time and Motion* (in Russian, Moscow, 1971), and other works.

BOUBOU HAMA. Writer and research author. Former President, National Assembly, Republic of the Niger.

Publications include: *L'Empire de Gao—Magie et Coutumes des Sonraïs* (co-author); *Enquête sur les Fondements et la Genèse de l'Unité Africaine*; *Kotia-Nima* (*Rencontre avec l'Europe*); *Merveilleuse Afrique*; *L'Aventure Extraordinaire de 'Bi Kado', Fils de Noir*; *Histoire Traditionnelle d'un Peuple, les Zarma-Songhay*; *Histoire du Gobir et de Sokoto*; *Recherche sur l'Histoire des Touareg Sahariens et Soudanais*; *Histoire des Songhay*; *Contribution à l'Histoire des Peul*; *Contes et Légendes du Niger*; *Sonni Ali Ber et Askia Mohammed Touré, Deux Bâtisseurs de l'Empire Songhay*; *Bagouma et Tiégouma, une Aventure où l'Afrique se Conte à Elle-même et à ses Enfants*; *Le Roman de la Sagesse Africaine*; *Réflexion sur l'Homme et son Destin*; several plays; (all published in France). *L'Histoire Traditionnelle du Dallol Bosso*; *L'Exode Rural, un Problème de Fond*; *Jeunesse et Développement*; (all published in the Republic of the Niger). Articles include: 'Traditions Orales et Structures de Pensée: Essai de Méthodologie', in *Cahiers d'Histoire Mondiale*; 'La Religion Africaine comme Essai et Source de la Civilisation de l'Oralité', in *Présence Africaine*.

HANS-GEORG GADAMER. Born 1900, Marburg, Lahn, Germany. Ph.D. in Philosophy. Professor of Philosophy, Leipzig University (1939); Dean of the Faculty (1945); Rector of the University (1946–47); succeeded Karl Jaspers at Heidelberg (1949). Member of various Academies (Leipzig, Heidelberg, Darmstadt, Athens, Rome, Boston). Awarded the *Pour le mérite* in 1971. Co-editor or correspondent for various German and foreign learned journals.

Publications in English include: *Truth and Method*; *Hegel's Dialectic*; *Philosophical Hermeneutics*. In German: *Plato und der Dichter*; *Goethe und die Philosophie*; *Über die Urprünglichkeit der Philosophie*; *Platos dialektische Ethik*; *Kleine Schriften: I. Philosophie. Hermeneutik*; *II. Interpretationen; III. Idee und Sprache, Platon, Husserl, Heidegger; IV. Variationen*; *Die Aktualität des Schönen.*

LOUIS GARDET. Christian philosopher. Specializes in comparative religion and cultures (particularly Muslim). Numerous visits to Islamic countries (including the Near and Middle East, the Maghreb, India and Pakistan). Visiting professor at the universities of Rabat, Algiers, Cairo, Beirut. Editor (with Etienne Gilson) of *Études Musulmanes*, Paris. Current contributor to the *Encyclopédie de l'Islam.*

Publications include: *La Pensée Religieuse d'Avicenne*; *La Cité Musulmane, Vie Sociale et Politique*; *Expériences Mystiques en Terres non Chrétiennes*; *Connaître l'Islam*; *Les Grands Problèmes de la Théologie Musulmane*; *Dieu et la Destinée de l'Homme*; *L'Islam, Religion et Communauté*; *La Mystique*; *Études de Philosophie et de Mystique Comparées*; *Les Homme de l'Islam*; and, with G. C. Anawati, *Introduction à la Théologie Musulmane—Essai de Théologie Comparée*; *Mystique Musulmane—Aspects et Tendances, Expériences et Techniques.*

AHMED HASNAOUI. Born 1949 in Sfax, Tunisia. Graduated from École Normale Supérieure, Saint Cloud, France. *Agrégé* in Philosophy. Teaches Philosophy at the University of Paris-X, France.

Publications include: 'L'Islam: La Conquête, le Pouvoir' in *Histoire des Idéologies* (ed. F. Châtelet).

TED HONDERICH. Reader in Philosophy, University College London. Visiting Professor at Yale University, the Graduate Center of the City University of New York, and Brooklyn College, United States.

Publications: *Punishment, the Supposed Justification*; *Three Essays on Political Violence*; various articles on logical theory (e.g. 'Truth' in *Studies in Logical Theory*, 'On the Theory of Descriptions' in the *Proceedings of the Aristotelian Society*), political philosophy (e.g. 'The Worth of J. S. Mill's *On Liberty*', in *Political Studies*, and 'The Use of the Basic Proposition of a Theory of Justice' in *Mind*), and the philosophy of mind and moral philosophy ('A Conspectus of Determinism' in *Proceedings of the Aristotelian Society*). Editor of *Essays on Freedom of Action* and *Social Ends and Political Means*. General editor of *The International Library of Philosophy and Scientific Method* and *The Arguments of the Philosophies*.

ABEL JEANNIÈRE. Born 1921 in Saint-Paul-en-Pareds, France. Doctorate on the pre-Socratic Philosophers. Professor of Political Philosophy, Institut d'Études Sociales, Paris, and of the History of Political Ideas at the Law and Economics Faculty, Beirut, Lebanon. Deputy Director of *Projet* (Paris), a monthly review of political, economic and social problems.

Publications include: *La Pensée d'Héraclite d'Éphèse* in the *Philosophie de l'Esprit* series; *The Anthropology of Sex*; *Espace Mobile et Temps Incertains* (co-author).

SAUL KARSZ. Born 1936 in Buenos Aires, Argentina. Professor at the National University of Buenos Aires, the Superior Institute for Professorship and the Institute of Contemporary Studies. Since 1967, teaches at the René Déscartes University (Paris V) and Paris VII University.

Publications include: *Lectura de Althusser*; two contributions to *Truth and Non Violence—A Unesco Symposium on Gandhi*; *Pédagogie des Sciences Sociales*; *Théorie et Politique*: *Louis Althusser* (German translation published, Italian and Spanish in preparation); *Sociologie pour Travailleurs Sociaux* (in preparation).

SEIZO OHE. Born in Kobe, Japan, 1905. Graduated from Tokyo Imperial University. Further studies in philosophy at Heidelberg, Germany. Professor of Philosophy at [Nihon] Nippon University, 1949–75. Fulbright Exchange Scholar and Rockefeller Research Fellow at Chicago and Harvard Universities, 1955–56, and Fellow at the Center for Advanced Studies in Behavioral Sciences, Stanford, 1958–59. Litt.D. from Kyushu University, Japan, 1959.

Member of managing and editorial board of Japan Association for Philosophy of Science; President of the Philosophy of Science Society, Japan; Councillor, Asiatic Society of Japan. Member, International Institute of Philosophy (Paris); Board of Directors, Council for the Study of Mankind (Chicago).

Publications: contributions to *Les Grands Courants de la Pensée Mondiale Contemporaine*, *Philosophy in the Mid-Century*, *La Méthode Axiomatique dans les Mécaniques Classiques et Nouvelles* (*Les Grands Problèmes des Sciences*, 13), *Contemporary Philosophy*.

PAUL RICŒUR. Born in Valence, France, in 1913. *Agrégé* in Philosophy, doctorate in literature. Worked at the Centre National de la Recherche Scientifique. Taught History of Philosophy at Strasbourg. Former Dean of the Faculté des Lettres of Nanterre University, France. Now Professor of Philosophy at the University of Paris-X and at the University of Chicago, United States. Director of the *Revue de Métaphysique et de Morale*.

Publications include: *Philosophie de la Volonté*: I: *Le Volontaire et l'Involontaire*, II: *L'Homme Faillible*, III: *La Symbolique du Mal*; *Histoire et Vérité*; *Platon et Aristote*; *De l'Interprétation, Essai sur Freud*; *Le Conflit des Interprétations*; *La Métaphore Vive*.

ARNOLD TOYNBEE. Born 1889, died 1975. Educated at Winchester College and Balliol College, Oxford, United Kingdom. Fellow and tutor, Balliol College, 1912–15; Government service, 1915–19; Member of British Delegation, Paris Peace Conferences, 1919 and 1946. Professor of Byzantine and Modern Greek Studies, University of London, 1919–24; Director of Studies, Royal Institute of International Affairs, 1925–56.

Published works include: *A Survey of International Affairs*; *A Study of History*; *Hannibal's Legacy*; *Acquaintances*; *Experiences*.

JOHANNES WITT-HANSEN. Born 1908, Levring, Denmark. M.A. and Dr Philosophy, University of Copenhagen, Denmark. Professor of Philosophy, University of Copenhagen. Member of the International Institute of Philosophy; World Future Studies Federation; Danish Social Research Council, Commission for Future Studies; and the Gottfried Wilhelm Leibniz Society. Awards: Mary Taylor Williams Fellowship in Philosophy, University of North Carolina, United States.

Publications include: *Exposition and Critique of the Conceptions of Eddington Concerning the Philosophy of Physical Science* (dissertation) and *Historical Materialism—The Method, the Theories*, Book I (both in Danish). Editor-in-Chief of Danish edition of Karl Marx's *Capital*; *Grundrisse der Kritik der politischen Ökonomie*; and *Theories of Surplus-value*.

CONTENTS

INTRODUCTION

Paul Ricœur

The thirteen essays collected here confront the reader straight away with a strange paradox. The diversity—inevitable, even desirable—of the viewpoints does not result only from the multiple cultural backgrounds of the authors, the variety of their specializations and interests, their differences of philosophical outlook. I was really fascinated by a deeper paradox. How can the same work encompass two such radically opposed concerns as that of the analytical philosopher, intent on reducing what we say of time to a conceptual minimum, and that of the thinker and meditator, intent on imparting to our experience of time its spiritual maximum? Yes, how, in speaking about time, say in one instance the least, in another the most? Not that the two are in contradiction; rather they really do not meet. One is interested, as we noted, in 'what we say about' time, the other in the 'spiritual maximum' of our experience of time, a question of minimum and maximum, conceptual economy or spiritual intensity.

Our sole aim in these introductory pages is to help the reader to think about this paradox and understand its necessity. To my mind, it derives from a fundamental feature of our experience of time, namely, that time is never lived directly, that it is never a mute, immediate lived experience but one that is always structured by symbolic systems of varying complexity. Some of these systems come logically and chronologically first and are immanent in different cultures; others are built upon the first through reflexion at a second degree, through philosophies, religions and popular wisdom. It is these symbolic systems which our authors explore, ranging between the extreme poles of a conceptual minimum and a spiritual maximum.

The rational skeleton of time

An analysis limited to identifying 'temporal relations and temporal properties' (see Ted Honderich) *ipso facto* prevents itself from inquiring into the many ways in which we valorize our experience of time. Such conceptual asceticism could be termed pre-symbolic. This is not to deem this type of analysis somehow inferior to the meditation of the thinker or sage. Instead, it raises a question considered as preceding any other investigation concerning the sense we attribute to what we say about time. As we do speak about time, it is legitimate to inquire into meanings without which there could be no discussion of time. Here Ted Honderich's article is particularly striking in carrying conceptual asceticism one step beyond the point where an analysis of ordinary language would normally seem to stop. What, in fact, do we find in our everyday language about time? Terms such as 'before', 'after', 'during'; but also such terms as 'present' ('now'), 'past' ('yesterday'), 'future' ('tomorrow'). The former involve relations only: relations of anteriority, posteriority, or simultaneity between any given events. The latter, involving temporal properties, add a reference to a particular event—the present, which exists now, and in relation to which all other events have been or will be present. Relations and properties provide an intelligible skeleton for our discussion of time. Conceptual analysis means first isolating both from the rest of our discussion of time and then delimiting each in relation to the other. Thus, the relations of anteriority, posteriority, and simultaneity between events do not change and are always true, while the present does change, the same event being in turn future, present, past. But the author's 'conceptual asceticism' reduces properties to relations, treating 'before', 'after', and 'during' as being alone irreducible. This reduction (the principle only is indicated here and the reader should refer to the article itself) rests on the following argument: to say that an event is now taking place is to say that it is simultaneous with the linguistic event which states it. The linguistic event in turn implies a consciousness for which statements can exist. The author terms his analysis 'cautious'—as distinct from an 'affirmative' analysis which sees something specific and irreducible in the experience of 'now'. But, in the author's view, this recourse to an opaque and supposedly primary experience can be interpreted as an acknowledgement of the failure of the analysis itself: 'What is affirmed is principally one conviction about the present, and it is affirmed in spite of the fact that nothing of an analytical kind can be said about it.' On the other hand, nothing remains opaque if what we understand by now is the simultaneity between an event that is happening and the event of our talking about and being conscious of it.

The 'cautious' analysis is perhaps irrefutable as such. But it reveals by contrast what we also mean when we speak about time, something that is implicitly contained in the notion of a temporal property. With simple relations (before, during, after) we are still dealing only with unspecified events or rather with possible events. In this sense, nothing happens. It is only with the allegedly opaque now that the experience of time commences. And precisely as temporal, this experience unfolds only in conjunction with the experience which Plotinus called the *diastasis* and Augustine the *distentio* of the soul and which consists in the split, the distance by which each new present, as it arises, pushes away the recent present which, in its turn, recedes into the past, while at the same time continuing to be retained by the new present which separates from it. Without this distance, this distension of the soul, there would be no lived experience of time. For nothing would then be anterior, posterior or simultaneous to anything else. In truth, nothing would happen. What permits the analysis to begin would itself be lacking—namely, *present* events—as would what permits the analysis to continue—namely, a real consciousness which produces the present utterances in relation to which other events are deemed contemporaneous.

'Cautious' analysis, therefore, has two effects. It does what it sets out to do, i.e. to analyse what we mean by temporal relations and temporal properties. It also shows up the gaps left to be filled by other approaches to time, readily admitting that 'cautious' analysis lacks persuasiveness, and that affirmative analysis is closer to our perceptive beliefs and pre-philosophical expectations. But why do we persist in thinking that we mean more by what we say about time than what analysis authorizes us to affirm?—no doubt because our perceptive beliefs and pre-philosophical expectations call for other treatment than sticking to the minimal meanings involved in what we say about time.

It is these other approaches that are purveyed by the symbolic systems by which we try to make sense of an experience which analytical philosophy has quite correctly shown us to be opaque. It is because it is opaque that it can be expressed only in symbolic systems whose cultural articulation is unavoidably manifold, divergent, even contradictory.

But before entering the labyrinth of symbolisms by which we attempt to articulate our own experience of time, a second starting-point must be considered, pre-symbolic in that the type of analysis it involves does not call upon cultural interpretations either but again limits itself to describing the rational skeleton of the signification of time.

The first starting-point was formal, dealing only with relations (before, after, during). The second is material, primarily concerned with our apprehension of objects subjected to change. We speak of time, says Yakov F. Askin ('The Philosophic Concept of Time'), 'when we refer to a sequence of events or to their duration'. It may be noted that this second type of analysis characterizes Marxist thinking about time, while the first more or less typifies Anglo-Saxon analytical philosophy. For the materialist philosopher, time has an 'objective reality' conferred upon it by matter in motion. Geochronology indicates that material entities record the passage of time; biology refers to 'biological clocks'; and psycho-physiology analyses conditioned reflexes as time functions. While it may be true that consciousness alone recognizes the passage of time and organizes time around the present, this present is always more than a simple point. Its weight derives from the persistence of certain situations and certain states. The relative stability and discontinuity or discreteness of material phenomena underpin experience of the present. And it is the *development* of real phenomena that connects past, present and future, that makes the passing of time one-directional and irreversible, and orients all levels of reality towards the future—matter, life the human individual, history. *Development* also involves tendencies, so that nature can be described in terms of real but unfulfilled potentialities, leaving room for innovations within the limits set by the laws of nature.

This future-directed orientation therefore does not stem from man or from human consciousness. Anticipation ('extrapolatory') reflexes studied in birds suggest rather that the psychic reflects objective reality, an anticipatory reflection rooted in materiality itself. Similarly, cybernetics is precise about the rational analogy between human behaviour and that of self-regulating systems: 'This tendency [towards the future] is the expression of a quality which does not pertain to human consciousness alone, but in a sense informs the whole process of development in the material world'. In all its forms, man's capacity for coping with the future (making plans, scientific forecasting, rationally organizing society, safeguarding the environment) proceeds from the same orientation towards the future which derives from the *development* of the material world.

It is understandable that in this materialist conception eternity signifies no more than the continuous existence of the material world, the absence of a beginning and end of motion: 'inherent in eternity are the two basic features of time: duration and sequence'.

It will not escape the reader's attention that this analysis is the precise counterpart of its predecessor, which started from what we *say* about time and derived temporal properties from temporal relations (before, during, after), while the second starts from changes and developments we *observe* in reality. The first started from ordinary language submitted to conceptual analysis; the second is a generalization from the most recent research results. However, the two starting-points are not altogether strangers. The first analysis, in order to have relations, needs events. The second, taking real sequences of events, must take account of the relational aspect of time. From the viewpoint which contemporary science suggests, Askin writes, time is neither 'a separate "thing", in its own right, [n]or an independent process acting as a demiurge of reality'. In this sense, Leibniz has the better of Newton: there is no 'absolute time', time is a form of existence, a real connexion.

Apart from the fact that one starts from what is said and the other from what is real, both analyses strive to limit themselves to what is fundamental. They are both concerned to isolate out the conceptual minimum in our understanding of time. Accordingly it is not surprising to find that each leaves aside the more specifically psychological and cultural aspects of the experience of time. We saw this in the opaque character of the present in the first analysis. Similarly with 'superstructures' in the second, i.e. we move from the real but unfulfilled potentialities of material reality to man's capacity to create the future via the lived experience of time which the other studies in this collection describe from the perspective of their diverse cultures. 'To move into the future', Askin acknowledges, 'is not at all the same as to travel to a constellation which already exists, but has not been reached. To move into the future is to create the future.' Whence the specific features mentioned above of man's orientation towards the future; whence also the positive features of eternity. Eternity does not signify solely the absence of a beginning and an end: 'Eternity as the permanence of existence reveals the creativity of the process of time'. Whence also the concern to enhance our apperception of time, to valorize time. Whence, finally, the proud affirmation which concludes the essay: 'there is a place in this process [the movement of the real world] for man—man the inventor, man the creator.'

After the essays by Yakov F. Askin and Ted Honderich, a consideration of the symbolic structures of our cultural experience of time seems doubly necessary: first, to move from a formal analysis of temporal relations to the live experience of the opaque present and, second, in order to move from a material analysis of change to man's promethean claim as inventor and creator.

The symbolic structures of our cultural experience of time

The expression 'symbolic structures' has here the sense used by the anthropologist Clifford Geertz in *The Interpretation of Cultures.* Cultural groups as studied by anthropologists, historians, sociologists and psychologists are structured by norms, rules, representations, beliefs which form a matrix through which each culture interprets its own existence. No society, as such, directly and immediately confronts its own lived experience, but 'reads' it in the light of its particular cultural codes, which operate for man as genetic codes do for animals, models through which experience is viewed and organized. At the human level, these models function at once as 'models of' and as 'models for'. They reflect and they ordain, never one without the other. This is why each cultural code is always both the broad expression of the society's economic, social and political life, and a compendium of directives for change. As expressions or reflections ('models of'), these symbolic systems, employing rhetorical resources and speech, provide a justifying ideology; as directives for change ('models for'), they engender prophecy, eschatology and utopias.

We thus can see how appropriate is the expression 'symbolic systems', since 'models of' and 'models for' reflect the same relations between code and message, and between deep and surface structures, as other semiotic systems (as was implied above when we said that each culture uses symbolic structures to 'read' and interpret its own existence). Varying somewhat the analogy with language, we might add that the anthropologist (in the broadest sense of the term) interprets social existence at one remove in that he attempts to write down the symbolic systems of peoples of particular cultures, or rewrites their reading of their own social life.

The semiotic status of symbolic systems is confirmed by the place of works of language in a culture's interpretation of itself. To consider, more specifically, the experience of time, it is evident that no culture can refer back to its own conception of time without having resort to the vital 'narrative' activity which, linguistically, is expressed in an immense variety of stories, i.e. of narrative discourse. Myths which recount how things began, the marvels which nurture legends and folk tales, epics or tragedies retailing the deeds of heroes stronger and more intelligent than mortal men; chronicles, annals, tales on their way to becoming more conventional history, stories of men like ourselves, realistic fiction leading in modern times to the naturalistic novel—all these narrative forms show that man becomes aware of what we earlier called temporal properties or the potentialities of natural development only by interpreting them through narrative. By this I do not mean that a culture's symbolic systems find expression only in narrative form, but only that the universality of narrative (is there any culture that has no stories—and stories of immense variety—how many sorts of them really are there?) shows how symbolic man's consciousness of time is. In telling stories, men get a grip on their experience of time, find their way through the chaos of potential paths of development, use intrigues and their solutions to plot the all too complicated course of real human actions. In this way, man the narrator makes sense of the inconstancy of human affairs, which the wise men of many cultures have contrasted with the immutable order of the stars.

If we recognize the symbolic structure of our temporal experience, we must recognize and respect the diversity of the symbolic systems which organize this experience. They are like languages. Mankind did not develop on the basis of a single language. We know of no historical era with all men having the same language. This is why attempts to invent a universal language that would fulfil the same role in communication as natural languages (natural only in contrast to this hypothetical artificial language) have always failed. Similarly with time symbolizations—we know of no historical era when all mankind might have identically systematized temporal experience. This is not to say that the diversity of symbolic systems concerning time coincides with the diversity of languages (it was shown in the first book in this series that there is no one-to-one relation between the grammatical structure of a language and what is said in it about time. What a given language says is not mechanically determined by its lexical and syntactical forms). It is rather that these systems constitute specific cultural codes which sometimes coincide with major linguistic groups, and sometimes cut across linguistic or cultural groups and form subgroups within them.

This book, indeed, indicates how various are the interpretations of time, starting from the spontaneous and pre-philosophical level of cultural life. The authors have different ways of describing this diversity. Societies can be contrasted in relation to their relative evolution towards development (Honorat Aguessy). Boubou Hama examines a particular culture (Songhai) and constructs an ideal type (animism) without reference to a particular geocultural area. Saul Karsz, on the contrary, considers the geopolitical area of Latin America and the internal competition of several cultural models. Yet again (Louis Gardet and Ahmed Hasnaoui), we can use philosophical elaboration (what I referred to above as second-order symbolization) to reveal conceptual cleavages of a new type, involving both a global difference (as between Arab-Muslim and Graeco-Latin Western thought) and doctrinal oppositions internally deriving from processes of differentiation and conceptual elaboration. Finally, Abel Jeannière's description of the pathology of the experience of time in advanced industrial societies shows how fragile and subject to distortion the process of symbolization is in these societies.

Aguessy's essay, 'Sociological Interpretations of Time and Pathology of time in Developing Countries', provides a warning to anyone embarking on any consideration of the various approaches to collective time. He cautions the Western reader against the really ethnocentric temptation of considering the cultural expressions of advanced industrial societies as being universally valid (the question is raised below whether they share a single, canonical experience of time or whether their cultural edifice is not in fact a house divided against itself). In any case, idealizing his own culture, Western man tends to measure all other societies by their degree of variation from his own, measured solely in terms of industrialization. Having thus attributed a special and augmented value to the criteria of rapid change, of the superiority of what is acquired over what is handed down, of rational forecasting and programming, Western man is inclined to define by default the status in regard to time of developing societies. He will describe them as living from day to day, or archaic, or even consider their behaviour *pathological*. In so doing, he will simply have naïvely passed judgement on his own culture, which hankers after the past in protest against certain unbearable features of modern life. He will also have failed to notice the differences in the rhythm of life in cultures he has lumped together as being behind the times. Nowhere, in actual fact, do we find societies wholly uniform in their attitude to time; variation occurs not only between but inside cultures. If this is so, it is a mere illusion to wish to 'situate the collective times of different cultures in relation to a homogeneous time and a history with a specific orientation'. The modern-versus-tradition argument is simply this illusion at its worst.

To give my own thoughts on the subject, I should say that the illusion of homogeneous time and a supposedly oriented history derive from a failure to recognize the symbolic manner in which we interpret our experience of time. For it to be possible to obtain a perspective on all cultural attitudes to time, there would have to be a non-interpreted, non-symbolized universal time. But time cannot be apprehended in this way. The only conceivable universality is in the opening of each culture to all the others, exchanges between them as equals, each acting on and being acted upon by every other. (We shall return to this idea of Aguessy's later.) The specifically Western illusion is thus based on an abstraction, the abstraction of calculated time in economics-centred societies. It forgets the abstraction, the separation and isolation from the cultural totality; the Westerner also fails to see that the priority to economics is in itself an interpretation, an axiological choice, and hence a form of symbolization.

Boubou Hama's essay ('The Soothsayer'), can, like Aguessy's, be read as a protest against the claim to universality which goes with the Western perception of industrial time. It is also of special interest in two other respects.

First, it warns against the opposite illusion to that of a universal social time, namely, the illusion of a pre-symbolic time. The animist conception of time is certainly based on life forms that are close to natural rhythms. Nevertheless, it is impossible without some form of symbolism to express time lived in rhythm with the seasons, animal life, the body's breath, or in relation to the extremes of illness or imminent danger, to patience or expectations, or to man's place in what can be called a 'spiritual concrete' universe. The symbolic structures of such fundamental experiences as these are to be sought in stories and legends, in the *griot*'s recitation, in the soothsayer's utterances, in proverbs and in numerical esoterism, and in the practical symbolism of initiation rites and magic. Hence, no experience entirely lacks a symbolic vehicle, verbal or non-verbal. To use Wittgenstein's terms, language games and forms of life are in continual correspondence.

The second particular point of interest in the essay on animism lies in its attempt to take animism to the level of a philosophical 'conception' of time. To do so the author employs a method which I have taken the liberty of describing earlier in Weberian terms as the elaboration of an ideal type—animism—on the basis of a model experience—the Songhai culture.

There is something paradoxical about this elaboration; investigating the paradox will help to advance our thinking on the symbolic constitution of time. The paradox is this: how can the animist interpretation be communicated in our Western languages without using the categories of Western thought? Must not the discussion of animism in languages structured by other attitudes to time be necessarily misdirected? How can we speak of the indistinguishability of matter and spirit, of the junction of infinite time and space, of the material and spiritual totality of the universe, in languages which have made a point of distinguishing between these and at times of setting them in opposition? In other instances, we need to separate where Western languages make no distinction, for example, between God and the demiurge who watches over the material and spiritual universe. We have even to dare speak of material time and of intemporal time. Transcribing the categories of one culture into those of another necessarily results in certain oddities; to make itself understood with the resources of a different symbolic system, a culture may be forced to use its own categories quasi ironically, making them say something other than that for which they were intended. But this, surely, is the paradox of all translation. When there are radical cultural differences, translation is both necessary and impossible: necessary because each of us, ultimately, understands himself only by comparison; impossible, because translating animist categories, for example, into the language of a Western philosophy deriving from Greek thought involves distorting and subverting one without ever doing full justice to the intentions of the other. And yet, there being inevitably a plurality of symbolic universes, translation remains the impossible but necessary task entailed by what, as we have seen, Aguessy regards as the only conceivable universality.

Within a single geopolitical area—Latin America—Saul Karsz ('Time and its Secret in Latin America') investigates variations on 'the psychological and moral valuation of time', a reasonable undertaking if it is true that time is never simply lived but evaluated, i.e. symbolized and interpreted. Apart from its wealth of information about Latin America, the essay is notable for its manner of varying the line of attack in relation to different parameters. The author first attempts to classify experience of time in terms of racial groupings: Indian, Spanish (Portuguese), Black, Mestizo, Creole. This is no biological definition, however; each group seems to be constituted (or rather established) by its way of reading its own time: colonial time as temporary uprooting, the former colonized peoples' time as time run out, the Creole's time lacking any sort of depth, and so on. Groups think about each other's time, and this interplay affects the symbolization of social time. Karsz, however, also considers time in relation to a system of political classification somewhat as Karl Mannheim does in *Ideology and Utopia*. The major currents of political thought can, in fact, be fairly readily distinguished in terms of their attitude towards time: the time of the Enlightenment and universal reason, the profound time of the Romantics, the industrious time of positivism, the impetuous time of Bolivar and Martí militancy, the time of challenging revolutionary movements, and so on. Karsz follows the same line in literature; he suggests that the artist, by raising symbolization to its highest degree of lucidity, reveals time's 'secret' (the word appears in his title): the writer revives the buried past, delves into the paradoxes and mysteries of a 'history of eternity' (Borges), or meditates on the overlapping of the times of mediaeval and of modern man in the many-layered time of Latin American man. Cosmological time can be brought to the level of historical time only through value judgements which are always partial and conflicting. An evolving continent like Latin America furnishes the best possible model of what might also be called the evolving symbol.

Like the preceding essay, the study 'Certain Notions of Time in Arab-Muslim Philosophy' (Ahmed Hasnaoui), is both historically informative and an invitation to consider anew the differences between symbolic systems. The author himself suggests this view by siting his contribution as a protest against the 'meta-position' of an (incidentally recent) imperialistic approach to history, the assumption by the West of its authority to speak comprehensively both for itself and for all others. As a branch of knowledge, the alleged supremacy of history as so defined appears in epistemological debates arising out of the newer human sciences. Basing himself on Heidegger's research on transcendental ontology and on Foucault's archaeology of knowledge, Ahmed Hasnaoui uses his reflections on Arab-Muslim philosophy to help deflate this monopolizing pretension. If the ontological tradition of the West is governed, as Heidegger says, by the pre-understanding of being as presence; and if, according to this pre-understanding, the essence of time lies in the fundamental correspondence between the instant, the point, the limit, and the absolute here and now; can the work of the great Arab-Muslim speculative thinkers then be included in this ontological tradition?

Having stated his problem in these terms, the author has a double inquiry to make: first, what weight does Arab-Muslim philosophy attach to what Heidegger referred to as the 'unthought' in Western metaphysics?; secondly, by what power of differentiation do these philosophers undermine the initial model? Thus it is shown how many of Aristotle's problematical and even paradoxical concepts (e.g. the analogy between 'now' and a point) are grounded in a particular tradition, and also how a different type of conceptualization—or rather a plurality of conceptual universes—comes into play in rendering an account of what the author calls 'the experience lying at the root of the Arab concept of time'. This, he says,

> can only be found in this ethico-cosmological perception of *dahr*, an unending recurrence 'of nights and periods'. the 'alternation of night and day', the substance of things, the element in which they move and are fulfilled, but also an anonymous power, the irreversible development of which carries things and beings in its wake and, as Imru'u-l-Qays says, 'permits nothing on its path to exist independently'. *Dahr* is also the place of testing, of *sabr*, the Arabs' quintessential ethical experience. *Ṣabr*, in spite of what is often said, is not indifference to events but rather the ability to cope with events. This experience should be understood in the sense of the 'ethos', that is to say, of a mode of behaviour which is not yet reflective or philosophical but which is a way of dwelling in the world and entering into interhuman relationships.

This fine passage seems to suggest that it is possible to grasp in itself, without intermediaries, the experience lying at the root of the Arab concept of time. As the author says: 'In order to grasp the experience of *dahr* one must go back to its origins prior to its philosophical codification by Ibn Sīnā and before Rāzī elevated time to the level of an eternal principle.' But he then concentrates on showing how this radical experience operates behind diverse, singular and conflicting conceptualizations.

He first approaches the 'temporal models whereby the experience of the community is codified' by way of a discussion of the theological instantaneism attributed to Arab thought by Louis Massignon, showing that this experience of the community can be apprehended only by making a linguistic effort to reappropriate these models.

For the author, their unavoidable diversity is, moreover, the consequence of what he calls the 'dilemma' posed by the fundamental phenomenon of the end of prophethood, 'the "seal" put on further prophecy'. For there are several possible attitudes in the face of this dilemma. And the discourses which are constructed on the basis of these attitudes become more and more complex. The doctrine as related by Maimonides, to which long passages are devoted, thus takes place among the 'secondary elaborations'; similarly with Kindī, founder of *falsafa* and author of the proof by *reductio ad absurdum* of the finite nature of time, endowed with a beginning and an end. Here, we are far from the original, founding experience.

This is not yet all. With Sufism a mystical strain appears. The reference event is now no longer the limiting instant but the reality of time experienced in the flash of revelation and in the sudden decision to renounce the world. Man no longer delimits but is delimited in the unexpected encounter with truth. Consequently, we have to balance the affirmation that a culture as rich and complex as Arab-Muslim culture has as its yardstick a primitive experience and the same author's affirmation that 'in Muslim thought, time takes many different forms'.

Louis Gardet's article, 'The Prophet', which we shall refer to later in speaking of the role of different key figures in the symbolization of time (prophet, leader and so on) decisively confirms this unitary yet plural reading of the Muslim conception of time. Unity is sought here in the central experience of prophecy (considered for the moment independently of any comparative classification of prophets):

The mere notion of prophecy implies a very special relationship with time, a break in the inevitable time sequence, an irruption into daily life of the points where time and a world beyond time meet, a reading of the facts of history in a light which transcends them.

Indeed, in and through prophecy, the Word of God 'makes time and dominates it'. Yet, on the other hand, this core of prophecy, common to Judaism, Christianity and Islam, cannot be apprehended outside the specific system of interpretation of each of the religions concerned. Gardet accordingly goes on to differentiate each type of prophetic experience in accordance with each's attitude to time. The Islamic experience divides into different currents of thought whose history is followed until the synthesis with Hellenism made by the Eastern Falāsifa (Fārābī and Ibn Sīnā). The reader will note that the similarities and differences between our two Islamic authors are of less significance than their agreement that the most radical experience ('the easing of the weight of days in a moment of transcendent duration') becomes accessible to us only through diverse, complex and contradictory traditions.

To conclude this review of the problems raised by the symbolization of the experience of time, I should like, as a sort of argument *a contrario*, to use Abel Jeannière's analysis in his article on the pathology of time in modern societies. This, of course, was not the purpose of his article. It is very clearly addressed to the dramatic problems posed by the discord between the respective rhythms of personal and social development. But in the light of the preceding analyses, the pathogenic structures engendered by industrialization can be interpreted as destructuring the symbolic universes inherited from the past, i.e. acting as de-symbolizing agents. They thus provide negative evidence of the symbolizing through which we enter into possession of our experience of time.

In treating the pathogenic structures of time in modern societies as forms of de-symbolization, I do not mean to minimize the emotional impact of the phenomena Jeannière describes: segmented time, disjointed time, imposed time, time seen as a penalty, accelerated time, time without reference to the past, and so on—raw experiences which directly affect time-regarding processes at their most immediate. This pathology has to be described in terms of desire, fear, frustration and disappointment. But every pathogenic phenomenon leaves its mark of suffering on us because the interpretation matrices which enable us to apprehend our own experience have themselves been disjointed. The measures and constraints imposed by industrial labour and the resulting distortions traumatize us because we no longer have a model capable of harmonizing the new forms of social time with the old structuring symbols. An 'inertia of mental change' derives largely from this disharmony of implicit representations, and hence of the symbolic codification of experience. Some of the phenomena described by Jeannière are directly and explicitly produced at this level of representation. Everything said about the depreciation of reference to the past, the lack of consideration for the old, the flight towards a future that lacks the security of former values, the total preoccupation with the present, the drop-out phenomenon among the young, indicates that the very representation of time, both individual and collective, is pathological. It must be added that the remedy proposed by Jeannière under the heading 'The prospective approach' (which we shall discuss below) cannot, in turn, be conceived of without a certain process of re-symbolization. If we are indeed time-sick, this is precisely because we do not know how to re-symbolize our experience of time.

The proper use of time

The pathology affecting the experience of time in advanced industrial societies does more than negatively confirm the importance of symbolization in ordering that experience. It also shows that symbolization is incomplete, fragile, and continually under threat. This is what distinguishes cultural codes, impregnated with history, from genetic codes, which ensure the repetition of the same models in animal species. Because of the place they allow to history, the above analyses have made it evident that even the most traditional models of time experience are subjected to continual discussion and reinterpretation. The creative aspects of these discussions can now be examined as we come to consider the other contributions to the book.

As it happens, several authors connect up the formation or transformation of models with the personal initiative of some remarkable people whom they portray. This is highly instructive. One can hardly speak of models (in the twofold sense of 'models of' and 'models for') without also speaking of the figures who both exemplify them and affect them through their teaching, actions, or meditation. Even in the contributions already mentioned, the various attitudes to time were linked with human types who were both their bearers and begetters. The animist conception of time is conveyed by the language of tradition, the storyteller's words, the soothsayer, by elders and wise men—in short, by all those who hold spiritual power (Boubou Hama forcefully stresses this connexion with power and those who hold it). Similarly, the power of prophecy, which Gardet regards as a victory over the irreversibility of time, demands the voice of the prophet, the man elected to utter the word which is stronger than time. Prophecy is not a reading of time except when its witness (quite often its martyr—the same Greek word means both) gives it the.force of a creative utterance. With the soothsayer and the prophet, those mentioned by the other authors can be included: leader, futurologist, guru, Zen master and—why not?—the characteristically Western manner of philosophizing.

One remark more before continuing, to state two of my own preoccupations. The first I have just mentioned: I link models to human types or figures in order to stress how dynamic and active is the symbolization which makes human time. The second leads back to the question with which I began this introduction: how are we to understand the central paradox of an inquiry which moves between the two poles of a conceptual minimum and a spiritual maximum? I hope to show that, by moving from abstract models to concrete figures, meditation on human time can rise towards this spiritual maximum.

It is primarily the man of action, the leader, who shows that time is not only lived but, we might say, acted out. The leader, Toynbee declares, is someone who weds initiative and opportunity. The opportunity may be a crisis which calls for the decisive intervention of a charismatic leader; or a slow fermentation over a period of time which permits the historian to say subsequently that 'the time was ripe' for an 'historical initiative'. The 'great man' (military or spiritual) sees the crux of the crisis, knows what people await; but he is also ready to seize the opportunity when it offers. Missed opportunities, hasty undertakings, the hero's failure to play his role, negatively portray the structure of opportunity—*kairos* I should say—which was Hegel's subject in his great introduction to the *Philosophy of History* as he meditated on the 'ruses of reason' in history and the role played in it by great men.

The futurologist (Johannes Witt-Hansen, 'The Futurologist') is involved in a different type of intervention. He is not a 'great man'. His milieu is not that of opportunities which it needs extraordinary gifts to recognize, but the probable evolution of well-defined and well-analysed social phenomena. His vision of the future steers its course between determinism pure and simple and outright libertarianism, both of which he rejects with equal fervour. Johannes Witt-Hansen takes advantage of disagreements of Marxists themselves concerning the inevitability of the passage from capitalistic to socialist society to show that Marx himself was unable or did not choose to present his predictions about the coming socialist society in the form of a proof; in particular, the transformation rules he bases on the interest on capital are no more than suggested trends and conjectures. It can be deduced, in more general terms, from this critique of mechanistic interpretations of Marx that futurology will never be prophecy, i.e. will never read the future as if it were past. But it does not presuppose an undetermined future any more than an inevitable one. To underscore the limits which circumscribe anything man can do about the future, Witt-Hansen stresses the double circumscription to which social time is subject: to cosmological time, which itself conforms to the law of increasing entropy, the second law of thermodynamics; and to ecological time, which compels us to confine our expectations and dreams of change within the limits imposed by the finite energy resources of our planet. The futurologist's work is therefore of necessity coloured by 'the fatal race between our attempts to restrain our consumptive waste of available energy, and our attempt to acquire the information required for escaping our present global predicament'. The 'future time' of mankind is thus not just any imagined future. The notion of saving energy, against the background of increasing entropy, implies a radical critique of uncontrolled growth, of the idea of unlimited consumption and, ultimately, of the gospel of prosperity common, it would seem, to both capitalism and socialism.

Can the figure of the futurologist subsume all the others? The author does not go so far:

> It may still be true . . . that a specific methodology, i.e. a theory of forecasting in general and social forecasting in particular, is practically non-existent. A global existential philosophy, demanding a revolution in our way of thinking of human problems is so far in *statu nascendi.* We are confronted with the problem of unravelling the logic of man's 'future time', including the logic of inevitability and possibility, necessity and contingency, determinateness and choice—a study still in its preparatory phase.

Our earlier reflections help us to understand why this 'global existential philosophy' and this 'revolution in our way of thinking of human problems' are still only beginning. The futurologist is not dealing only with quantitative phenomena accessible to conjectural calculation but also with considerations of value which, as Jeannière's article shows, are affected by a specific pathology. Jeannière's 'prospective approach' (looked at here in conjunction with Witt-Hansen's 'global existential philosophy') is not simple. It must encourage ethical and educational allies to counter the distortions in the experience of time of our advanced industrial societies. In other words, it must add therapy to forecasting. Hence, as scientific forecasting of the future, futurology must join with thinkers whose resort is to the most profound wellsprings of wisdom.

The teachings of the Eastern and Far-Eastern masters of wisdom are represented in two essays. Sri Madhava Ashish introduces the guru, Seizo Ohe the Zen master. Each regards the master of wisdom as a master of eternity, and each believes that thought feeds on direct experience. The symbolism which organizes human time is in this sense transcended, superseded by something higher, and the experience of time advances towards its spiritual maximum. And yet, symbolism is not entirely abolished, in so far as the experience of eternity is transmitted by the master's discourse to his disciple and elaborated upon in books of wisdom. Spiritual experience is thus reflected in traditions as different from one another as, for example, the ancient wisdom of India differs from present-day Zen Buddhism.

The guru, then, can well be called a seer, saint or wise man, and be acknowledged as a numinous figure who reveals and exemplifies by existing. His oneness with other seers and with the wise men of all the other religions can be forcefully affirmed ('They have "seen" the one, timeless truth. . . . Different though their temporal forms may be, in themselves they are the same'.) The elevation of time to eternity, which they all teach, may well be warranted by an experience deriving from both life and thought. But the fact remains that the teaching relation disciple to master demands the exercise of critical thinking, so as to reject the principles of materialism, positivism, or scientism (as the author does at length in the polemical part of his article), or to distinguish the true guru from the charlatan. Above all, access to the vital experience is available only through the meditation on doctrines in which the ground common to all teachers and seers separates into different currents. It is as if, as soon as it is spoken, the unformed must take on form, and submit to the constraints that symbolization involves.

Sri Madhava Ashish distinguishes, for example, at least two major interpretations of the individual's identity with the universal: he must lose himself entirely and forever in universal bliss; or he may keep his spiritual life just this side of complete dissolution, out of compassion for the suffering of all men. Both certainly share the same shedding of self, the same relinquishing of all personal motive; but the

dual interpretation of the goal possibly accounts for there being two recognizably different attitudes taken by seers towards the world: those who sacrifice the bliss of self-annihilation commonly accept the bonds and responsibilities of affection here, while those drawn to annihilation commonly reject them.

Seizo Ohe, too, starts from a fundamental experience in which time is identified with the conscious self, and that self with all things: 'In each moment', says the sage Dōgen, 'there are all existences and all worlds'. Time, then, is no longer a simple passing through. The eternal present is contained in the here and now. This experience, in which time is all, is that spiritual maximum which I originally contrasted with the conceptual minimum which suffices for the formal analysis and the materialist description of time. Where analytical philosophy sees only relations between unspecified events, and where materialist philosophy concerns itself only with real changes, Zen wisdom describes an all-inclusive experience of time. For Zen, wisdom does not limit itself to analysing ordinary language or to extrapolating from results provided by science, but transcribes the illumination of the Buddha, the first Awakened, into thoughts and sayings. The path opened up by this Enlightenment resembles Kant's in considering time as that which orders the content of our experience and, secondly, that of Husserl and Heidegger, in seeing time unfolded through human existence and engendered *hic et nunc*. But the Buddhist path leads much farther, right on to the intemporal void of the eternal present. This place of nothingness is reached only when the desire which engenders sadness at time passing is itself extirpated. Then, it is not only daily cares that disappear but the illusion of the separate self.

Once again, however, differences between schools reappear. Seizo Ohe distinguishes between the passive attitude in which the ego gives itself up to the Power-of-the-Other and the active attitude of the *Zazen* who, through practical training and breathing exercises, cultivates the Power-to-be-Oneself. Seizo Ohe suggests that the development of the Zen tradition in contemporary Japan could, by re-establishing the lost harmony between body and soul, set man's original creativity free. In placing the emphasis on creativity in *Zazen*, the author implies that Zen, in its contemporary form, could respond to the needs of the industrial world mentioned above in connexion with Jeannière and Witt-Hansen: 'On the whole', says Seizo Ohe, 'science today stands surprisingly close to Zen; both respect experience and see reality as it is. To use Dōgen's phraseology, both are "studying the truth", but science is more "forgetting the self", while Zen is more "studying the self".' If this were true then the last link would be closed, making the joint between the spiritual maximum and the conceptual minimum.

But can the joint be made in this way, not only between science and Zen but, more generally, between logical thought and wisdom?

The West may still have something to say here, in counterpoint to the wisdom of the East and the Far East. This wisdom proclaims that time is consciousness and that consciousness is all things. To this inclusiveness, whose method is to annihilate all determinations in a timeless void, Hans-Georg Gadamer, in 'The Western View of the Inner Experience of Time and the Limits of Thought', speaking in this respect for one of the major currents of Western thought, replies that we cannot get away from the experience of time: 'The inborn awareness of death, the experience of youth and of ageing, the irrevocability of the past, the unpredictability of the future, the daily distribution of our time and planning of our work—all these involve time in one way or another.' Time is an experience which thought always runs up against. What thought cannot encompass, because the imagination is limitless, is the beginning and the end. And thought is rendered all the more incapable of thinking this unthinkable barrier by the practice of its own kind of asceticism towards myth and epic, its asceticism in conceptualization. And so it takes Saint Augustine's anguished plea as its own: 'What then is time? If nobody asks me, I know; but if I wish to explain it to someone who asks me, I do not know'. This unknowing becomes even more opaque when thought, reflecting on death, sees in the relation to death something specific to our human lives. Man alone is a creature so individual that he cannot understand his being on the model of the indeterminate alternation of life and death: it is death which imparts duration to his life.

This thought suffices by itself to refute the idea that cosmology alone governs our conception of time (as in Plato's *Timaeus*, where time comes into being along with the world, as the dimension immanent in the movement of the heavens). If, as we have read in Aristotle, time always presupposes a soul which distinguishes between instants and counts intervals, then our reflection on time must unceasingly be led from the consideration of the world to the *distentio animi* described by Augustine (mentioned above in connection with the distinction between 'temporal relations' and 'temporal properties'). This 'spiritual tension between distraction and concentration, fear, hope and repentance' is what makes it impossible to get away from the experience of time. In this regard, Heidegger does no more than carry Saint Augustine's intuition to its lucid ultimate when he makes time an 'ontological' moment in the structure of our life and when he makes the orientation towards death a feature of 'authentic existence'.

However, it would be to miss the point to conclude that this meditation on finiteness could eclipse the other ways of approaching the problem of time. Both change in natural things and a certain intemporality implied in the exercise of thought continue to force themselves concurrently upon us. The will to overstep all boundaries is also continually affirmed in the successive versions of the myth of Prometheus—the god and idol of modern civilization. And, again, the ecstasy of creativity and of love never ceases to evoke in us a certain idea of eternity.

None of these paths is closed, none is prohibited by the meditation on the finiteness of existence, since it could never replace all our other ideas about time without itself becoming an absolute. Its purpose is rather to set a limit to any ambition which aims at subsuming our many perspectives on time into a discipline which would itself transcend time. This limitation on knowledge, become the knowledge of limits, is perhaps the West's reply (or at least one of its replies) to the East (or to a certain East).

This reply is not, however, to be understood only as a retort; in one sense, this awareness of limits *co-responds* to the un-knowing which wisdom in the East, the Far East, and elsewhere, recognizes in regard to the experience of eternity.

The instant, the immediate,
the now—and eternity, the all, the one

The arc of time (past, present, future)

Temporality (phenomenological analysis)

THE WESTERN VIEW OF THE INNER EXPERIENCE OF TIME AND THE LIMITS OF THOUGHT

Hans-Georg Gadamer

One of the most inscrutable mysteries confronting man when he tries to reach an understanding of his own existence is that of the nature of time. The inborn awareness of death, the experience of youth and ageing, the irrevocability of the past, the unpredictability of the future, the daily distribution of our time and planning of our work—all these involve time in one way or another. The measurement of time was a revolutionary discovery no doubt made at a very early stage in the history of mankind, probably not at one inspired stroke but rather in a variety of circumstances and in different places and ways. The question of the nature of time must have arisen in a very similar way, being one of which each individual becomes aware as soon as he begins to reflect on his own life and experience. Such is the power of our mind and imagination that we can conceive the infinite; but in so doing we always come up against two unthinkable barriers: beginning and end. Herein lies the crux of the mystery of time: that although everything we encounter in reality is finite, our imagination knows no finite bounds which it cannot overstep.

Religion and myth offer an inexhaustible variety of answers which it would be fascinating to investigate. But however much mythic tradition reflects the riddles of ending and beginning, death and birth, nothingness and being, it does not provide a real answer to the problem of time. For myths do not treat the nature of time by means of concepts. Nor can we accept as a religious or mythological answer the interpretation of mythological or religious texts. For the conceptual tools with which the commentator approaches the text will have a decisive influence on the sense which is made of it. Ultimately this means that however much we recognize the profundity of statements passed down through the history of religion, there is only one way along which we can continue our investigations: our Western culture has opted for the way of the concept, which means the way of philosophy.

This much is clear from the meaning of the word 'philosophy' itself.

Originally, it covered a far broader field of reference than the narrow, specialized sense in which we use it today. The much wider original meaning of the Greek word is 'devotion to purely theoretical interests'. Philosophy therefore takes in the whole field of scientific knowledge, excluding only the practical or technical applications of such knowledge. Even after Plato had restricted the word's meaning, giving it the new and significant slant of man's thirst for knowledge which can never be quenched by the fullness of wisdom, this being reserved for the gods, the broader meaning of the word lived on. It is well known that Aristotle's *Metaphysics* was called the 'First Philosophy', which meant the first of the sciences. It was not until modern times, when with Galileo the new physics gave the word 'science' a special cachet, and only when the new ideal of scientific method gave modern science its own specific characteristics, that the word 'philosophy' began in its turn to take on the narrower meaning with which we are now familiar. Today philosophy is no longer the science among others, perhaps even the foremost, but a form of knowledge on which all other forms of knowledge are based, yet which at the same time aspires to be different from them in that its subject is 'being' in general. This wholeness of vision is only vouchsafed to the various branches of knowledge in fragmentary fashion as research becomes ever more specialized. The word 'philosophy' has yet another meaning, of course: the popular meaning of practical wisdom which it has had since the time of Socrates and which is to be found in a similar form in the practical philosophy of Aristotle.

From this etymological digression we learn that Greek thought laid the foundations for science when it rejected the colourful world of myth and the dramatic immediacy of the world vision contained in epic poetry. When science—which at the time consisted mainly of mathematics, astronomy and music (and to a certain extent also medicine, as the science of human nature)—set rational thought and abstract reasoning in the place of the mythological imagination, immediately the question of the absolute arose, requiring a rational answer. Only with the emergence of the desire for rational knowledge expressed in terms of concepts could such a question as ours—What is time?—be asked. But one could equally well reverse the order and say that it was the conscious posing of such questions that led to science and eventually to our modern civilization which is based on science and which has now spread to cover the whole world.

One thing is certain: it could not have been observation alone. Even if one considers the whole range of man's responses to time, his experience of the alternation of day and night, of the cycle of the seasons, the growth and withering of plants, the conservation of the species from one generation to the next, the transitoriness of each individual living creature; however conscious one is of birth and death as the confines within which the human drama is played out, however desirous of probing the mysteries of the hereafter—this all leads up to the question, 'What is time?', but does not answer it. Yes, the question of time seems more intractable and perplexing than all the others with which we as philosophers are familiar, for example the nature of substance, causality, matter and form or even space. It seems nonsensical to deny that there is such a thing as time, but whether it exists in the same way as other things which we describe as existent is open to doubt and objections of the utmost urgency. One is reminded of Augustine's famous words in Book XI of the *Confessions*: 'What then is time? If no one asks me, I know; if I wish to explain it to one that asketh, I know not.' But is he in this well-known formula really referring to the special mystery of time, or to a universal experience in philosophical thinking, the gap between the practical, pre-reflexive use of our concepts and our inability to define them? The philosopher's formulation of concepts (Hegel's 'travail of the Notion') is continually being bogged down in the eternal refractoriness of words (*to agèraton pathos tōn logōn*) in all the ambiguities, misunderstandings and antinomies which are almost inseparable from any kind of philosophical argument and which from Socrates' times to our own have given philosophy its characteristic tension. Unlike poetic, epic or mythological narration, where truth is not asked for but whose meaning is beyond question, philosophical abstraction is always open to the suspicion that the problems it deals with have no *fundamentum in re*, no basis in reality, but are merely ways of playing with words. In our century in particular, this criticism has frequently been levelled at the traditional 'metaphysical' approach. The trail was blazed by the logical empiricists and the analytical philosophers of the Oxford School, and this marked the beginning of a veritable revolution in the world of philosophy, as all the celebrated problems of metaphysics appeared to disintegrate in the light of linguistic analysis. The problem of time was no exception. J. N. Findlay felt that he had unmasked the problem of time in 1941 by placing an embargo on 'meaningless' expressions as Wittgenstein had done. However, in the 1956 reprint of his article Findlay seemed already to have reverted to a more positive evaluation of the philosophical problem of time.

All in all, the problem of time seems to be a particularly revealing one in philosophy. The different cultures and periods of history can be quite clearly distinguished one from another by their differing attitudes towards time. There is, for example, the well-known belief that Christianity and subsequent thinking broke with the Greek concept of cyclical time in favour of a rectilinear historical time. Indeed there is something plausible about the idea that, in contrast to the Greek concept of time modelled on the periodical rhythms of nature, a conception of time according to which time moves steadily towards an infinitely distant future is based on a prefigurement of the 'last days', a final state of everlasting bliss and eternal peace. In other words, the idea is that it is eschatological and theological motives which have led, in their secularized form, to the concept of history which sees the flow of time as unidimensional and irreversible, and that the concept of history and historical time which underlies modern thought simply did not exist in classical antiquity.

However, if one accepts this argument one might well wonder whether the view that Western civilization began with the Greeks is correct. Rather than concerning ourselves with the Greeks' efforts to think their way through the nature of time by means of concepts, should we not concentrate on Christian eschatology and its Utopian secularization? It is seemingly in relation to 'world history' and God's plan of salvation for the world that the concept of history and with it the reality of time first comes under the scrutiny of philosophy.

Once more we have something to learn from a word. This time it is the word 'modern' which provides the key to the mystery. The Ancients did not feel that the latest thing was automatically the best; indeed it would be true to say that for them history was equivalent to decline, the decline of an order of things which had proved its worth, its disintegration into the varying play of changing events. Does this mean that the Greeks had no feeling for history? Does it mean that we can set the more recent approach to time seen as a linear progression against the Greeks' endless repetition of cyclic movement which measures and numbers time in the same way as our hours?

The Greeks' alleged lack of feeling for history is supposedly exemplified by the story of Solon in Egypt. Plato tells how Solon identified ancient Egyptian stories of the flood which were described to him with the Greek traditions handed down in myth and how the Egyptian priest mocked such naïvety with the words 'You Greeks will always be children'. In fact this story is far from revealing a lack of feeling for history. On the contrary, it illustrates the density of the Greeks' awareness of history, for which myth and reliable traditional accounts of the past are knit together in the seamless robe of their national history. This is by no means indicative of a lack of historical feeling or complete absorption in the here and now of contemporary life.

A better illustration of the Greek attitude to time is provided by another Greek, Alcmaeon of Croton, the physician, who said: 'Men must die because they are unable to link their end to their beginning.' This saying is obviously intended to refer to the specific nature of man. It is however true of all living creatures—with the exception of the gods—that none escape death and link their end to their beginning. Life is always bound up with death. In referring to the linking of beginning and end, however, what the saying has in mind is not so much the special ontological status of the immortal gods, but rather the cyclical rhythms of nature, the regular return of sun and moon, spring and autumn, the pattern followed by all living species. But it then goes on to affirm: man differs from all other living creatures which like him are subject to the natural law of life through death, yet confer duration on life as such. Only man is so individual a creature that he does not lose his identity in the continuation of the species but is aware of his death, fears it and tries to hide from it. Man does not therefore accept the eternally renewed alternation of life and death as the pattern for his own being, and this makes him aware of his own powerlessness: he is set at variance with nature and with himself. Alcmaeon expressed this in the dictum 'All things human are two' and compiled a 'list of opposites'. For him, the distinguishing feature of man is precisely this duality and inner conflict which leads him into the hazardous adventure of thought.

Profound though it is, it is true that Alcmaeon's saying does not yet pose in explicit terms the question of the nature of time. This will come later, in the first complete work of Greek thought, Plato's *Dialogues*. Here, particularly in the *Timaeus*, the subejct of time is raised, in the context of the creation of the universe by the mysterious craftsman who in so many ways resembles God the Creator and yet, despite all his majesty and superiority, remains strangely different from the God of the Old and New Testaments.

The riddle of time is not approached here as the object of a question regarding the nature of things, to which Socrates' interlocutor is unable to find the answer. Timaeus comes upon the question of time as if by chance, in the context of a mythical story. Nevertheless, this story forms the point of departure for all subsequent philosophical analyses of time. The first known discourse on time, in Aristotle's *Physics*, refers back to the 'definition' of time given in the *Timaeus*, and in its turn ushers in a lengthy tradition.

This tradition is always linked in some way with the question of whether such a thing as time really exists. When Augustine in his *Confessions* becomes engrossed by the riddle of time, the contradictions and perplexities which emerge do not admittedly raise any doubt as to the reality of time. However, even here such doubts are not far below the surface. The conclusion is that time eludes all understanding, so that we cannot truly say that time exists, except in the manner of striving towards nothingness, in other words as passing away. Does time exist? Or is it, as the philosophers say, merely a subjective form of the apprehension of reality? When, at the height of the Middle Ages, Scholasticism gave new life and vigour to the heritage of the Greek philosophers, doubts as to the reality of time which had remained beneath the surface of Greek thought broke cover once again. Indeed, the famous edict of 1277 with its list of condemned heresies includes a rejection of the statement that time exists only in apprehension, not in reality (*in aprehensione, non in re*). This suspicion is obviously deep-rooted in all human attitudes towards time: Indian and Chinese parallels are not hard to find. But what is so characteristic of Greek and, with it, European thought is the fact that the Greeks continued to ask questions about the nature of time, despite the confusion into which it threw them, and even in the face of the divine revelation of eternity. Western thinkers continued to regard the question of the nature of time as crucial, and refused to dismiss time as an illusion or mere phenomenon, regardless of the ontological perplexities in which it involved them. Indeed, it can be said that Greek ontology's failure to solve this problem was partly responsible for the increasingly refined and penetrating analyses on which the philosophers of the Christian era had to embark if they were to succeed in their task of transmitting the heritage handed down to them by the Greeks.

Seen from this angle, the mythological description in the *Timaeus* takes on an unexpected dimension. At first sight the story of the creation of time appears to be merely a mythological presentation of our experience of time, and in particular of our measurement of time. For the underlying idea is that it is the universe itself, in spite or because of the regularity of the motion of the celestial bodies, which first makes man aware of time. The changing positions of the constellations challenge the human spirit to discover the laws which govern all these variations, and this, in turn, will lead man on to the discovery of numbers and the measurement of time. But the deeper meaning becomes clear when one considers the everlasting oneness as taking the form of a huge spinning-top made up of all existent worlds, which continually revolves around us, its axis. Man's first attempts at star-watching are shrouded in the mists of time, predating by thousands of years Plato's question, What is time?, but they were indubitably a response to the challenge which the observed anomalies—presented in particular by the orbits of the planets—pose for human intelligence. However, before the emergence of Greek astronomy such activities consisted merely in the accumulation of observations in the hope of revealing the hidden rules underlying the irregularities. For example, astronomers would compile a table of the eclipses of the sun (*saroi*). It is typical of Greek thinking, however, that this practical counting and measurement of time and recording of the observed results no longer satisfied, but led on to the 'theoretical' question as to how the universe is really constituted behind its superficial image and what role time plays in this connexion.

Timaeus describes the creation of time by the demiurge as one more stage in the creation of the universe. When the demiurge had created the world's 'soul', that is to say had endowed it with its own power of movement—the ordered motion of the constellations around the earth—he wished to improve upon his handiwork still further. The visible world had to be an exact replica of the ideal universe, i.e. having the structure of a living organism and able to maintain itself in a perfectly regulated equilibrium. The motion of the replica must therefore also be governed by rules of the highest mathematical perfection, for that is how it is in heaven.

And now the story takes a most unexpected turn: in order to make this visibly revolving universe of stars even more like its ideal paradigm, the demiurge created time.

This ideal original on which the visible world and its temporality are modelled is itself assigned something resembling an eternal 'lifetime'. Like each mortal creature it has 'its' time. But this lifetime, which in Greek is called an aeon, has in this case no limits, for the ideal pattern as a regular mathematical order knows no change, motion or end. It is a continuing 'there' which time imitates in the form of movement, the structure of this imitation being that of the numbered Presents counted out in a numerical series. This is time. This is how time underlies the motion of the circling All and is inextricably bound up with its endless progression.

Obviously time is not identical with this movement of the universe, but merely accompanies it. It is a movement in which nothing in fact moves but whose 'being' is just like the ongoing infinity of numbers. The numerical series 'is' endless, even if this is something we do not know from experience. The same is true of the motion of the All 'in time', and time itself is quite independent of all that really has motion. Thus, without coinciding with the movement of the heavens, time prefigures its articulation. Time is thus one aspect of the structure of the world itself, an attribute by virture of which particular times, and whatever moves in time, can be measured.

Plato's mythological description of the establishment of time was completely reshaped by Aristotle in his analysis of the key concepts of physics. His own analysis of time does not really differ from the underlying meaning of the *Timaeus*, but adds a new sharpness of outline to its opaque metaphors. On the one hand, he relates the ontological aporias which lie beneath the surface of the problem of time to the paradox of the continually vanishing Now. On the other, he is the first to play on the connexion between the nature of time and the nature of the soul. Plato's indication in the *Timaeus* that determination of the role of time in the structure of the universe leads on to number was already a pointer in this direction. For what is the ontological status of number? Does it not have a special connexion with the soul, that is to say with human intelligence? How else but through the human spirit's power of abstraction can pure number have being? The degree of abstraction required matches that of the structure of time itself, which is unlike anything to be found 'in time'. Aristotle therefore raises a crucial point when he refers to the relationship of time with the soul, although such references are only episodic and do not compare with Augustine's handling of the subject, marked as this is by the pregnant character conferred by his introspective, existentialist standpoint.

The mythological account given in the *Timaeus* in itself leaves no room for doubt regarding the 'reality' of 'time'.

However, the ontological certainty of the reality of time is slightly eroded if we take a closer look at the Greek background against which the concept of time is set. What is the real meaning of the mythological narration of the *Timaeus*? There is one obvious contradiction at the very outset: the creation of time is referred to merely as one stage in the creative activity of the demiurge, not by any means his first exceptional achievement. Much else was done 'before' the creation of time. Does this not mean that the whole story calls for a much more metaphorical interpretation? Plato himself drops a hint to this effect in the words of Timaeus (*Timaeus* 34 *c*), who warns that his story is not completely free of coincidence and arbitrariness and emphasizes in a most uncharacteristic way the precedence of soul over body. There is another explicit statement to the effect that time was created 'at the same time' as the universe (38 *b* 6). It is hardly surprising that a violent controversy arose in Plato's Academy regarding the credence to be given to the whole story of the creation, some considering it to be merely a device to lend vividness to the description.

Apart from the obscurities and contradictions in the chronology of the god's creation of the world, the account of the creation of the world's soul, which immediately precedes that of time, contains some very unusual features. One must of course bear in mind that the word 'soul' in the original Greek sense of *psyche* does not in the first place imply consciousness or even the possession of senses: the primordial function of the soul is to 'inspire' in the sense of endowing with animation. This means that living organisms—and most obviously animals—are given the power of independent movement. Consciousness, on the other hand, which in man takes the form of understanding or intelligence, appears to come from a quite different source, as it were from outside (*thurathen*) according to Aristotle. As the Greeks thought of the universe as a single living creature, albeit a large one, this tells us something about its form. It must be spherical, as it embraces everything. It must have an inner life, as it is a living entity. But because nothing exists outside it, attributes such as sense organs would be unnecessary and meaningless. The Whole accordingly takes the form of a system if for the purpose of describing it we carry out the appropriate reductions when viewing it as a living creature. What is surprising, however, is that Plato, when describing this system of the All, in which soul and body are united, should refer to the 'circles' of sameness and otherness, obviously to be understood with reference to the plane of the ecliptic. But he then goes on to relate these to true discernment and correct opinion respectively, even going so far as to mention sensory perception in this context! Where does this soul suddenly appear from, manifesting the familiar forms of human cognition? Should we try to make sense of it or rather dismiss it as obscure mythological verbiage?

Let us try to investigate the objective relationships between such concepts as identity and difference and the structure of thought, speech and understanding. It is obvious that none of these forms of cognition could be envisaged if it were not for the existence of identity and difference.

This question is dealt with by Plato in the *Sophist*. The inspiration which he follows here in the *Timaeus* is to make continual play of the analogy between the structures of the world order and of thought. The regular ordering of the universe presupposes identity, e.g. the alternation of day and night. Where things are ordered in articulated fashion both are always to be found. On the other hand, we have the ability of the human mind to conceive such an order, that is to distinguish between difference and identity and yet to see them together. To imagine the opposite of a universe revolving in concentric circles about a single focal point as described above will make the point clear. There is a close relationship between difference and discrimination. Things which are different must be distinguishable. All discrimination and separation must allow the differences to emerge and this process refers back to the existence of a, so to speak, endless observer, for whom the process occurs. One could go even further and say that it is impossible even to describe the system of the universe without positing an ideal observation point, for which the Whole exists. In this way identification and discrimination can be seen as none other than the obverse of identity and difference themselves.

In the light of these considerations, it is easier to understand the emphasis with which Timaeus affirms (37 *c*): 'If anyone calls that in which opinion (*doxa*) and discernment (*episteme*) can alone take place anything but soul, he is speaking anything but the truth.' 'Being' and 'soul' are inseparable.

But then what about the additional quality of time, its close connexion with the articulation of the motion of the universe? Its numerical structure refers back to the same principles of sameness and otherness, oneness and duality. Plotinus was not perhaps so wrong when he opposed the Aristotelian derivation of time from motion and argued that 'soul' is the origin of both time and motion.

The *dramatis persona* of Plotinus which throws itself into time out of desire for self-realization is of course more like a gigantic metaphor for the endlessness of the numerical series and of ceaselessly onward-moving time, devouring her own children, the Nows, than a genuine reference to the human soul which numbers and scans the flux of time. It was the passionate self-examination of Augustine which for the first time cast on the ontological paradoxes of time—which can never 'be' in the present and yet is nothing but the succession of all these self-consuming Presents—the light of the inner experience which the soul has of itself as it presses onward to meet the future. This was Augustine's way around the ontological impasse into which we are drawn by asking the question: When is 'time' actually 'present'? He revealed the dimension of inwardness and shifted the focus of attention from time itself to man's awareness of it, finding the existential dimension of time in the spiritual tension (*distentio animi*) between distraction and concentration, fear, hope and repentance. But from the purely metaphysical angle, and in particular as one of the key concepts of physics, the concept of time remained henceforth inextricably bound up with the ontological contradictions which rear their heads around the nature of the present—until in the end the nature of time was either no longer a subject for discussion and time served only as a parameter for the measurement of motion (Newton) or else the 'numbered Presents' were referred to only in the context of the inner dimension of consciousness, i.e. of the inner life (Kant). Yet the continuing emphasis on the unidimensionality of the inner sense in which impressions succeed one another reflects the ancient ontological approach and the Aristotelian definition of time as the measurement of movement and succession; and the 'travail of the Notion' to conceive the 'being' of passing time and to elucidate its ontological structure reaches its culmination in Husserl's masterly analysis of time consciousness.

It is surely a fundamental human attitude rather than a mere consequence of Greek ontology to regard time as something to be used, counted and measured. This use of time implies the ability to conceive of time *in abstracto*, i.e. as 'empty', and this is what Aristotle has in mind when he speaks of the 'feeling for time' as characteristic of man. It is his distinguishing feature that he is able to sacrifice the immediately pleasurable for a greater good in the future. But it must be admitted that this is a very one-sided way of experiencing time: it is seen as being at man's disposal, as 'empty' time, a homogeneous expanse stretching before him into the distance as if specially smoothed out for his benefit. Scheler has made it clear that it is the outward-directed excess of drive which leads to the conception of time as 'empty'.

There are, however, other ways of experiencing time, in which its reality is construed not as something which we encounter only when we attempt to reckon with it but as something which becomes operational within human existence as an integral part of it. Ever since Heidegger made the ontological meaning of the temporality and historicity of human existence a new subject of discussion and drew a sharp distinction between this 'authentic' time and measured 'world-time', we have once more become aware of the constitutive role played by time as an 'ontological' moment in the structure of our life. Heidegger characterized temporality as an 'existential' of human existence, in the context of his new, methodical treatment of the question of being. The 'reality' of time emerges much more clearly, however, in other phenomena which we have in common with other living creatures. For childhood, youth, maturity, old age and death mark out each individual's path through life, and these milestones of the individual life are reflected in the institutions and customs of society. The experience which man acquires as he passes through these different stages is a genuine form of experience of time itself. It is not the same as the counting and use of time and cannot be tied in with the theories of time referred to above, which have their conceptual apotheosis in the Kantian principle of the ordering of phenomena according to succession and simultaneity. It is most closely connected with the historical sense, which is strictly speaking the awareness of epochs, of one's own epoch and, in an even more basic way, of the 'pastness' of an epoch: a stopping-place in the constant flow of time, the establishment of a 'block of time' consisting of the simultaneous or contemporaneous.

These other forms of time experience are not at all related to the constant flow of time, but have their basis in the organic unity of the living being. They represent qualitatively different kinds of 'time'. Even the individual's attitude to the end of his own life appears to be controlled by a structural law of time experience involving an element of experience of time and life shared by all human beings: the constant narrowing of the horizon of the future ahead of us. In like manner, as Hegel has convincingly depicted, there is a continual extension of the span of the past as it merges into the shadows behind us. For our awareness of life is governed by a wide range of heterogeneous factors in which both future and past are 'present': giving rise to hope and forgetfulness on the one hand and resignation and reminiscence on the other. This awareness of life is singularly bound up with the certainty of death, even when this is concealed and repressed. It was to this that Heidegger drew our attention when he affirmed that the 'authenticity' of existence lies in its being a 'preliminary to death'. On the other hand death is also firmly rooted in our awareness of our existence and in our assurance of life (as Max Scheler demonstrates in his posthumous papers, thus echoing the insights of many great thinkers and writers). Death is both the characteristic of man and the natural lot of all living things.

Man has found many different solutions to this awareness of death. Religion deploys its inexhaustible creative power in order to throw light on the obscure riddle of the transitoriness of all that is human. I am not qualified to assess the role of the Christian message in the history of the West. But we are obviously not only heirs to the Greeks' feeling for philosophy and their fascination with the world around them. Where the Christian gospel is accepted by believers, it is seen in the end by the philosophical observer as the most profound answer to the mystery of death. Death is truly overcome when it is seen in God Himself. This is the inspired thought at the heart of the mystery of God's Incarnation and His suffering and death for man on the Cross: God's death robs human death of its sting and overcomes the irreconcilable opposition between mortality and immortality. This thought was given a philosophical formulation by Hegel, when he placed the Resurrection and the outpouring of the Holy Spirit in the context of the philosopher's self-knowledge. The fusion of the Greek concept of reason with the Christian doctrine of the Holy Spirit, as attempted in Christian spiritualism and Hegelian idealism, sets a seal on all the speculative endeavour directed to the mystery of the Trinity which pervaded Western thought from Augustine to Hegel, and unites the legacy of the Greeks with that of Christendom in our culture. For the philosopher the question thus arises whether the specific characteristics of Western *experience of time* are not the result of a similar fusion of Greek and Christian thought and whether the contradiction between Greek worldliness and concentration on the present and Christian spirituality and concentration on the future is really as marked as the gap which yawns between Greek metaphysics and Christian doctrine.

But if in the course of Western history itself the Judaeo-Christian religious tradition fused in this way with the Humanist religion of self-improvement, Hegel's bold attempt at philosophical synthesis, culminating in the philosophy of absolute knowledge, is seen to be not so much daring as historically justified.

In any event, one must always bear in mind that the conception of time as organic has played a decisive role in all philosophies of history. In a way this can even be said of Hesiod and his famous description of the declining ages. His is not simply a tragic view of world history, in which everything is decay and ruin, as it ultimately points to a time of stability, even if the toils and troubles of human life represent a marked decline from that Golden Age. And Hesiod looks forward to a final equilibration of human life which will arrest the decline, by virtue of the fact that the order of Zeus holds sway.

The doctrine of the ages plays a similar role in Christendom and also, in a secularized form, in the eschatologies of the philosophy of history which lay a new stress on the future. What we find here is always a division of historical time according to its content, and not the homogeneous stream of passing time and the meaningless counting of moments leading monotonously on to the Day of Judgement or the final end of history. The end of time is also the fulfilment of time. This is the line which was also followed by men such as Vico, Herder, Fichte, Hegel and his successors. Such romantic poets as Novalis and Hölderlin and theosophical thinkers such as von Baader and Schelling in their own way defended the concept of organic time against the time concept of Newtonian physics and Kantian philosophy, and this view has regained currency in our own century, particularly since Bergson. He distinguished authentic experience of time from the time concept of the natural sciences, a theme which has in the meantime become a burning question in metaphysics, thanks to Heidegger's interpretation of being in the context of time.

But this has not, of course, ended the supremacy of the Greek concept of time, and in particular of the view that its main function is measurement. This aspect of the problem of time is too closely bound up with the Western awareness of existence, which has forged for itself a valid, even if increasingly disquieting, means of expression in the natural sciences and their technical applications. Time as something at our disposal when we plan, work and build is a conception which corresponds only too well to the empty succession of moments which was characteristic of the Greeks' idea of time.

The relationship between religious and philosophical attitudes to time, and in the same way between Greek and Christian attitudes, is thus one of the greatest complexity, which is not surprising as time itself has so many different aspects. The self-knowledge of the West and the special characteristics which have imprinted themselves upon it can therefore perhaps best be summed up by way of conclusion in considering our subject in the light of the changing fortunes of a symbol which itself has many different layers of meaning. This I think we have in the figure of Prometheus, one of the most significant of the Greek myths, a strange mirror, which alike reflects our civilization's progress towards science and our present cultural level. The survival of this mythical figure through the millennia is significant enough in itself. In my view, the twin pillars on which the great arch of this story rests are, on the one hand, the drama of Aeschylus and, on the other, Goethe's creative reformulation of it in his poetic and dramatic fragments. For Aeschylus, Prometheus is the hero of technical invention, while in Goethe—in the ode of that name, in the Prometheus fragment and in the (also unfinished) *Return of Pandora*—he is presented as a creative genius, Goethe no doubt using his imaginative gifts partly to interpret the poetic intentions of Aeschylus as best he could from the only surviving play of the trilogy.

There can be no doubt that Aeschylus' *Prometheus Bound* formed part of a trilogy. It seems to me quite inconceivable that this play as such was intended as Aeschylus' last word on the Greek stage. For Zeus, the father of gods and men, appears here as a brutish tyrant unable to control himself and there are references to his possible fall. The trilogy must have led the story to a satisfactory conclusion, to the reconciliation of Zeus and Prometheus, the admission of the Titans to Olympus, the official introduction of the cult of Prometheus, the god of potters, in Athens, and in general to the firm establishment of law and order under the reformed leadership of Zeus. This seems to me clear from the facts of the cult of Prometheus, unless one disputes the play's authorship and tries, as some have done, to ascribe it to the Sophists.

Within the framework of a trilogy, it was a stroke of genius by Aeschylus to make the mythical friend of mankind, who brought it fire from heaven, the hero of human civilization. He saw herein the possibility of a dramatic situation rich in tragic irony: the first to encourage man to help himself suffers helplessly at the hands of a vengeful Zeus! When Aeschylus relates the Prometheus myth to mankind's ability to use fire—and thereby acquire technical knowledge—and the long history of intellectual and technical invention in general, he places a new and most significant slant on the myth's traditional connotation and, in particular, on its relationship to the story of Pandora's box. The spirit of self-help inevitably leads to a reassessment of the value of hope. Hesiod's pessimistic description of hope, which alone remained in Pandora's box, as the only evil on which men can always fall back and which can always be relied upon to disappoint them is replaced by a new interpretation. Hope is no longer 'empty', an automatic response arising from despair and idleness, as it were the hallmark of the peasant, but the source of optimism and a new confidence, that belongs to the artisan! The gift of fire leads to inventiveness, manual dexterity and the endless march of progress and civilization.

In the *Prometheus* of Aeschylus this is also represented as a new attitude to time. There was once a time, when mankind was still in its primitive state, when each individual stared unseeingly towards the end of his own life, 'seeing and yet not seeing, hearing and yet not hearing'. But time is now an empty space to be filled by planning, activity and progress. For behind the veil which hides the hour of our death from each of us, there stretches out, before each of us, a path leading on into the future. And conversely it is this future which lends itself to man's powers of imagination precisely because the hour of his death is hidden from him. Western culture recognizes itself in both of these interpretations.

The description given by Aeschylus contains no reference to a threat hanging over mankind, whose fate does not appear to be affected by the relations of the gods between themselves. We must presume the final reconciliation of Prometheus and Zeus and the establishment of a new, stable world order to take place in the world of the gods alone. There is no allusion to mankind's proneness to self-destruction through wars and violence, no attempt to avert such catastrophes by extolling righteousness and godliness, no education through rhetoric (Protagoras) or philosophy (Plato). We must assume that both gods, Zeus as well as Prometheus, will have learnt something. And yet the account of events on Mount Olympus is to some extent a mythological reflection of the process of humanization and a representation of the tragedy of culture. Men too have something to learn. The question is: Have they really learnt anything? Will they ever learn? And what have they to learn?

The tragic figure of Prometheus has accompanied Western culture through the ages. The benefactor of mankind condemned to endless suffering could even be seen as an anticipation of the crucified Saviour. This can be illustrated rather neatly by the appearance in Christian Byzantium of a work entitled *Christus Patiens* which consisted of two thousand lines of Aeschylus' *Prometheus* without a single alteration or addition! An early Humanist joke, no doubt, but symbolic of the Christian era and its genius for assimilating external factors.

In later antiquity Prometheus was characterized not by his suffering but by his skill. He was the creator of man, the architect of the human race (in conjunction with the local cult of the potter god), a model which has been an inspiration, particularly for Western thinking, since the Renaissance. The creativity of the artist, who brings a new world into being through the force of his imagination, was henceforth related back to the Prometheus myth. The artist is a second god (*alter deus bovillus*) and Prometheus is no longer the fire-stealer but rather Aeschylus' founder of civilization, for in the final analysis he symbolizes creativity in general and not only that of the artist. The poet and mankind in general are alike in that both see in the independence of the Titan an example to be followed.

But how is one to interpret the new self-confidence of the Renaissance which asks: Are we gods ourselves? Does this show any awareness of the limitations of the human mind and will? Is Prometheus the god of modern civilization? Or would it be truer to say its idol?

A very far-sighted answer to these questions can be found in Goethe. He was a young lawyer, just twenty-four years old, when he wrote a poem and a dramatic fragment which gave new life to the figure of Prometheus, making him a symbol of the self-reliance of the creative mind. The plot of the drama can be outlined as follows: Prometheus builds his own world, gives life, with the help of Minerva, symbolizing wisdom and philosophy, to the creatures of his imagination, and raises up—and educates—a new race, the race of men. Pandora, the fateful gift of a jealous Zeus in the old myth, is in this version the innocent daughter of Prometheus and no longer represents any threat to the well-being of mankind. The self-confidence of Prometheus appears to be unassailable and he purposely cuts himself off from the society of Mount Olympus. He recognizes no master other than omnipotent time and eternal fate, to which the Olympian gods are just as much subject as he. This was the major insight of the mature Prometheus, that none of the Olympians was in a position to master time, which must obviously mean to master the future and the past by condensing them in the moment and the present. The fact that they cannot do this means that their eternity is nothing but endless duration, which is not in Prometheus' eyes anything to be particularly proud of. Goethe, who was nicknamed Prometheus by his friends in Strasbourg, appears to have felt that the creative self-awareness of each and every man embraces a like awareness of eternity. For it is true that as long as this self-awareness lasts, nothing can separate the individual from himself. This means that he exists in the unshakable conviction of his continuing existence and future. This is the deeper secret of self-awareness which is expressed in the words of Goethe's Prometheus: 'I am eternal because I am.' The thought that it is possible that this thinking self-awareness may one day cease to exist is in a way unthinkable. It is as if this feeling of impossibility sets an indissoluble seal on the transcendence which is part of our feeling for life.

But now the drama begins. Prometheus' self-assurance is not all. An overwhelming presentiment of something that exists on the other side of consciousness is brought to him by his own daughter. There is a marvellously poetic scene in which Pandora comes rushing to her father in a state of great excitement. Without really understanding it she has observed a love scene between her girl friend and a youth. When she asks him what was the strange feeling which she saw in her girl friend and herself experienced, Prometheus answers: 'It was death.' What an answer! When Pandora next inquires what death is, he answers by describing in glowing terms the kind of ecstatic experience in which everything clear and definite dissolves and in which self-awareness reaches the highest peak of self-absorption. This description of ecstasy will be readily understood as that of the climax of love and surrender. But once more Prometheus calls it death, not love. And what he describes as the next state, sleep, awakening and complete renewal, is indeed very similar to death. The figure of Prometheus here takes on a new dimension. Nothing would have led one to expect such bewildering experiences.

Goethe's dramatic work comes no nearer than the *Prometheus Bound* of Aeschylus to giving a clear and unambiguous answer to the question of what the message of the Prometheus figure really is. A reference to Goethe's autobiography, *Dichtung und Wahrheit*, can, however, throw some light on this subject. Here he states quite categorically that the character of Prometheus had become a focal point in his thoughts, symbolizing his own experience that in life's critical moments man can rely only on his own strength and personal gifts. For Goethe, creativity was the basis of his existence and he saw in Prometheus a symbol both of the solitude and independence of the creative genius and of the self-restraint which all men must learn to develop. It was not at all the Titans' rebellion against the gods which appealed to his way of thought, as he himself stressed later. Indeed, the drama of the self-reliant Titan remained unfinished and in later years, in his play about Pandora, Goethe portrayed the figure of Prometheus in a more critical light, contrasting the basic one-sidedness of Prometheus' practical proficiency with deeper experiences of life: love, death and the Dionysian miracle of resurrection. We may perhaps be allowed to interpret Goethe's insights. It is precisely these tragic limitations on the creativity of the Titans which make clear the hopelessness of self-awareness; it is in the experience of the other, and of the otherness of death, that he finds his limit. In the eyes of Prometheus, each passion by which one is overcome represents a separation from oneself, a loss of sovereignty and of autonomy. In Goethe's eyes, however, and in those of his hypothetical reader a new fellow-feeling and a new order of communal living and communal bliss must triumph over the exertions of the Titans, bringing with it the benediction of the truly divine. The tragedy of culture ends in the celebration of Nature and its final reconciliation with culture.

CERTAIN NOTIONS OF TIME IN ARAB-MUSLIM PHILOSOPHY

Ahmed Hasnaoui

The spectrum of our concept of time, which impinges on several areas of our thinking, is becoming ever narrower as one of its constituent elements —the historical—increasingly tends to overshadow all others; as a result of this, our view of time is now dominated by this obsession with history which may be said to be the philosophical malady of our age.

The concept of history is said to have arisen in the West once the Western world had adopted, with regard to itself and to other cultures, an attitude which was essentially a meta-attitude, providing a means of reflecting simultaneously on Western and other cultures. Such a meta-attitude is actualized, for example, in a discipline like ethnology and in what is termed the 'historical initiative'.[1] The dominance of history affects scholarship not only because of the development of the humanities and through the methodological discussions to which they have given rise but also as a result of our practical interest which requires the choice of a historical concept compatible with our political praxis. This process of choice often amounts to rejecting those concepts (such as evolutionism, mechanism, etc.) which are regarded as mistaken. Furthermore, the development of epistemology—particularly historical epistemology which is the form it has taken, especially in France—has also done much to establish the dominance of history.

The fact that we are mainly concerned with historical time is illustrated by our very mode of philosophical reflection which focuses less on themes that by definition belong to the field of metaphysics than on the indeterminate complex of ways of thought and behaviour that arise both in the human sciences and in the world around us.

Philosophical activity also takes the form of an account of the past history of philosophy, a dismantling of the interlocking pieces of the philosophical edifice. And we often find the concept of time running like a unifying thread throughout this process of dismemberment. Thus in

Heidegger's view ontology, from the ancients right down to Hegel and even Bergson, is predetermined by its implicit understanding of being as a presence, *ousia*, *parousia*. According to this view, the same attributes of time are to be found in the *Physics* of Aristotle and in the philosophy of nature expounded in Hegel's Jena *Logic*:

Aristotle sees the essence of time in the *nūn*, Hegel in the now (*Jetzt*). Aristotle conceives of the *nūn* as *horos*, Hegel considers the now as 'limit' (*Grenze*). Aristotle understands the *nūn* as *stigmē*, Hegel interprets the now as a point. Aristotle describes the *nūn* as *todé ti*, Hegel refers to the now as the 'absolute this' (*das absolute Dieses*). Aristotle follows tradition in relating *chronos* to *sphaira*, Hegel insists on the circular path (*Kreislauf*) of time ....[2]

Should Kindī,[3] Ġazālī,[4] Ibn Rušd,[5] Ibn Sīnā[6] be regarded as belonging to this ontological tradition in that they recognized the 'fundamental correspondence ... between *nūn*, *horos*, *stigmē*, *todé ti*'? Gazālī writes:

The discontinuous quantity differs from the continuous quantity: between the parts of a discontinuous quantity there is no part in common linking the extremities (*taraf: oros*) [. . .], in the same way as the point (*nuqṭa*)—the common (*muštarakā*) and ideal (*mawhūma:* imagined) limit situated in the middle of the line—links its extremities (or those of its segments), or the now (*ān*)[7] links the extremities of time past and time yet to come.[8]

This quotation implies that the analogous nature of the 'point' and the 'now' is self-evident and beyond doubt. The problematical manner in which it is propounded in *Physics*, Book IV is, so to say, forgotten and concealed. Indeed, although Aristotle emphasizes the similar character of the point and the line—both of them being a common and ideal limit—he nevertheless draws a distinction between them by observing that the points of a line exist simultaneously, whereas the 'nows' destroy each other and present themselves as a succession.[9]

The text which we have quoted seems to gloss over the asymmetrical aspect of the point/line and now/time analogy. It disregards the entire connotative setting comprising such other correlatives as potentiality/actuality, essence/accident, etc.[10] Does this denote the culmination of a tradition which crystallizes the reference text and, in place of its tentative and ever-renewed quest—often intended merely to identify and formulate the problems[11]—substitutes uncompromising and dogmatic statements? This would seem to be borne out by the fact that the immediate designation of time is not made in the terms of Aristotle's *Physics*[12] but in the more formal, more didactic terminology of the *Categories*. Time is stated unequivocally to be a *continuous quantity*.[13]

In fact, this designation is neither fortuitous nor is it merely the result of a hidebound tradition; it is an indication of the way in which this tradition is formed and of the role it is given to play in the conceptual strategies of the moment.

Thus the interrelationship between all those properties which, for Heidegger, indicate an ontological precomprehension of the being of time and a precomprehension of being in general might be merely the consequence of accepting the limitations imposed as a result of choosing the 'hypothesis of continuity'.

The subsumption of time in the category of the now may appear to be a transposition of the time-phenomenon as apprehended by a 'naïve' consciousness; however, the determining factor is not so much this apparently intuitive grasp of experience as the 'formalizing' process whereby it is defined in retrospect as a 'point', a 'limit', etc., thus relating time to movement and magnitude, the continuous nature of which has been affirmed. This 'choice' which Aristotle made in order to meet the problems which had arisen in Greek mathematics (the discovery of irrational numbers, speculation with regard to infinitesimals)[14] lies behind the characterization of the point as the *limit* between the segments of a line—a limit which is merely an indicator of the infinite divisibility of magnitude—and of the instant as the limit between the past and the future. The ideal character of the instant and of the point (time and the line no longer being composed of indivisible elements) follows from a decision which was entirely compatible with the state which Greek mathematics had reached, and provides an adequate explanation of all the distinctive features of 'formalized' time subsumed in the category of the now.

Aristotle thus became the referrent of a standard of rationality serving as a yardstick for other theses concerning the nature of time. Aristotelianism in general came to be regarded as the highest form of rationality, a fact which conferred on *falsafa* the twofold merit of, on the one hand, claiming the ability to adduce, at every stage in its argument, the totality of its grounds—which it can subsume under a name—and, on the other, aspiring to classify each discipline by specifying the type of discourse to which it belongs—rhetorical, dialectical, apodictical or demonstrative—thus embracing and categorizing all fields of learning. *Falsafa* is thus cast as the sole fountain-head of canonicity.

Thus, in his polemical works,[15] Ibn Rušd usually adopts a relatively complex attitude with regard to his adversaries (the theologians, Ġazālī, etc.); instead of rebutting their theses by adducing the truth which would dispel their errors, he gives full rein to the point of view he is contesting, to its logical conclusions, allowing it to follow through until it dissolves in a welter of absurdities and contradictions.

In *Manāhij-al-'adilla*,[16] for example, in order to demonstrate the groundlessness of the *Aš'aritic* hypothesis of atoms, he shows the inability of one discipline—theology (*kalām*), which he regards as being restricted to Rhetoric—to determine the nature of bodily substances. For this would require that a solution be found to such difficulties as the distinction between the continuous and the discontinuous and the relationship between the continuous and the infinite, which are outside the scope of *kalām*. This discipline is thus ascribed limits which it can exceed only at the risk of making what may be termed a 'transcendent' use of rhetoric.

Reversible or cumulative time

What we can nowadays criticize as the raw concept of history (in the sense of the unfolding of a homogeneous time continuum) is often said to be a prerequisite for the emergence of historical consciousness. The emancipation from the cyclic view of time—characteristic of the Hellenic approach[17]—was brought about by Christianity which introduced a linear conception of time.[18] Islamic time, on the other hand, is described by Louis Massignon,[19] for example as reversible time, to which 'temporal models' implying the idea of progress are inapplicable.

We need do no more than mention at this point that the idea of progress is not entirely alien to Muslim thought, and has not been so since the dawning of Muslim philosophy. Kindī, at the opening of his *Epistle on the First Philosophy*, praises

> those who put the fruits of their studies at our disposal and who smoothe our path towards that knowledge of true and hidden things which we desire by offering us the premises which give us easier access to truth. For if they had not existed it would have been impossible for us, even had we laboured unremittingly throughout the whole of our age (by which is meant the Arab period of history), to gather together these primary truths which enable us to reach firm conclusions.[20]

Kindī does not situate the emergence of truth in the remote past which is for ever lost or to which a return should be made and, unlike the Platonists, he does not conceive of knowledge as a kind of recall, but he dwells on the necessity of deriving present benefit from the heritage of the ancients (even those belonging to a 'distant race'),[21] which is only conceivable within the framework of *cumulative time* marked by continuous progress. Time appears not only as the guardian and abode of truth but as the 'place' where it grows, so to speak, nourished by new inferences drawn from premises which themselves have been accumulated over the ages. This law of the accumulation of knowledge applies to the whole of the period separating the indisputable premises bequeathed by the ancients from the new and as yet unknown conclusions which will undoubtedly crown 'our painstaking research, our detailed investigation, our diligence'. The quantity of truth is 'proportional' to the time spent in acquiring it.[22]

Thus, short of arbitrarily emphasizing certain of their individual pronouncements on the matter, there is no way of arriving at a single formula which epitomizes the Arabs' concept of time and embodies its secret. Moreover, it is not at all clear what meaning still attaches to reversible time when we set it beside the concept that has most strikingly characterized the Arab perception of time, namely that of *dahr*, as an indefinite and inexorable undulation which, passing over the places where the tribe of the beloved has settled, and effacing even the trace of ashes, elicited the poet's lament: There is no 'refuge' to be found where all vestiges have been swept away.[23]

The experience lying at the root of the Arab concept of time can only be found in this ethico-cosmological perception of *dahr*, an unending recurrence 'of nights and periods',[24] the 'alternation of night and day',[25] the substance of things, the element in which they move and are fulfilled,[26] but also an anonymous power, the irreversible development of which carries things and beings in its wake and, as Imru'u-l-Qays says, 'permits nothing on its path to exist independently'.[27] *Dahr* is also the place of testing, of *ṣabr*, the Arabs' quintessential ethical experience. *Ṣabr*, in spite of what is often said, is not indifference to events but rather the ability to cope with events. This experience should be understood in the sense of the 'ethos', that is to say, of a mode of behaviour which is not yet reflective or philosophical but which is a way of dwelling in the world and entering into interhuman relationships. *Dahr* is not only the ethic of the Arabs but also their aesthetic: is it not said that someone in the face of adversity 'is embellished' on putting on the cloak of *ṣabr*?[28] It is the authentic mode of sojourn in the 'sea of becoming', in so far as the latter may be said to be a place of sojourn. It is the ethical standpoint in the face of a becoming, in which the question of the free will of an autonomous agent does not arise.[29]

The 'original' Arab experience of time, like that of the Greeks in classical times,[30] is essentially ethical and cosmological, and *ṣabr*, in particular, can only be understood once it is re-integrated with this experience as an attitude of open-minded equilibrium facing the *hubris* of amorphous becoming.

In order to grasp the experience of *dahr* one must go back to its origins prior to its philosophical codification by Ibn Sīnā[31] and before Rāzī[32] elevated time to the level of an eternal principle.

The *mīṯāq* or pure Event

Just as Arab thinking about time cannot be circumscribed within the formula of reversible time, it cannot either be reduced to the terms of a theologically derived instantaneism because the opposite reading can also be made, at least at the level of the Shī'ite 'consciousness', the time-attitude of which is an expectation of the coming of the *Mahdī*, as of a goal or *telos* indefinitely postponed, which implies the unfolding of a continuous temporal line.

Louis Massignon has stressed the *discontinuous* nature of Islamic time, drawing attention to its links with an intransigent monotheism and an occasionalism which makes God the sole efficient cause. This apprehension of time is described by Massignon as absolutely general because it is referred to the most intimate form of dealings with the Divine as well as to the corpus of ritual practices, to theology, law, medicine and mysticism. And we can invoke grammar to bear witness to the absence—originally, at least—of the threefold dimension of time. This conception of time is, in Massignon's view, so pregnant in meaning that some explanation is needed of the manner in which juridical universals come into existence.

This manner of being-in-time is, according to him, a sort of concrete a priori of Muslim consciousness and knowledge which, because of a radical discontinuity necessitated by the rigid monotheism of Islam, are obliged to create, not without difficulty, something resembling duration as a prerequisite for the genesis and continued existence of the juridical universals, the statutes (*aḥkām*).[33]

But, apart from the fact that this analysis assumes an unbroken continuity between the Muslim 'world of life' and phenomena in the realm of ideas (juridical, for example), and is thus based on a misunderstanding of the specific mode of constitution of theoretical structures, it seems to us that the temporal models whereby the experience of the community is codified are at the same time more complex and subject to a different set of rules than that of a patchwork derived from instantaneism.

First, Islam has undermined the notion of *dahr*: 'Labour in this world as though you were to live eternally, labour for the next world as though you were to die tomorrow'. This pronouncement by the prophet, situating the 'moral' agent at the crossroads of a time whose end is the common round of humdrum activities and of a time constantly overcast by the shadow of death, defines the present as the privileged sphere of the ethos and thus, in effect, destroys the infinite and massive *aiōn* of *dahr*.

Second, in the *mīṯāq* we have a myth that grounds the specifically religious experience of time—the first creation in 'pre-eternity' of the entire race of Adam, that it may proclaim God the One Lord.[34] Although this is a *pure Event* in that it cannot be situated in time, it must however find its temporal (or spatio-temporal) realization in the central figures of prophethood each of whom expounds the *mīṯāq* anew in the idiom of his own day. This raises the formidable problem of the end of prophethood, to which there are two possible attitudes.

On the one hand the community may be regarded as being in full possession of the truth; in such a case the task is one of embodying it in a tradition or, in other words, in the recollection of vestiges, in the preservation of monuments, in the detailed recording of words and deeds which lay the basis for the philological and annalistic 'sciences', etc. This is the significance of the strict adherence to the letter of positive religion (*šarī'a*).

This is also the meaning of the various returns which are made to the roots of this tradition,[35] whenever it seems to be threatened by what is seen as an alienating form of rationality. Such figures as Aš'arī, Ġazālī and Ibn Taymiyya, in calling for such a return, have always sought to undermine this type of rationality by bringing sceptical arguments to bear on it.[36]

The literal purity of tradition as it was originally formulated is to be maintained, the original source and the letter being mutually reinforcing. Any departure from the source is a 'falling-away', a distortion of the meaning: hence the condemnation of reprehensible innovation, *bid'a*,[37] which is frowned upon in that it is in danger of regarding itself as a source and considering itself (or being considered) as the inauguration of a new tradition, valid as such for the community. *Bid'a* is a faulty, distorted replica of the original source, and as such reprehensible.[38]

In addition to these faulty replicas, there are true (commendable) ones, which consist in explicating the tradition and developing its content, hence the importance of *qiyās* (reasoning by analogy) which, in the *fiqh* (science of revealed law), resolves the problem of the finite, limited nature of the revelation.

This is the main mode in which the experiences of the community are situated in time; it is less a matter of painstakingly weaving a tissue of duration between two independent instants than of replicating invariant attitudes. This formal framework is so pregnant with significance that *falsafa* repeats it in its relationship with the corpus of Greek philosophy. Is not the reaction associated with Ibn Rušd the philosophical equivalent of the action undertaken by the *salafia* with regard to the corpus of revelation?

The second possible attitude to the end of prophethood is that espoused by Shī'ism and, in general, by those who reject the tyranny of the letter of revelation, namely the *mu'awwilūn*. Despite the differences between them, this attitude is governed by the same logical considerations as the first. In the dilemma posed by the end of prophethood—on the one hand it is necessary to repeat the pure Event represented by the *mīṯāq*, on the other, such repetition is in fact precluded by the 'seal' put on further prophecy—the Imamate represents a continuation and renewal of the original covenant.

In this way, a secret time, that of the succession of Imams (guides), invests empirical temporality. Of course the scriptural revelation necessarily remains the touchstone (since it is the meaning of the sacred text that is being sought), but as far as the community is concerned it is not so much a question of rooting it in a tradition as of 'pluralizing' it among a multiplicity of groups each of which is characterized by the position which it adopts with regard to one of the multiple levels of meaning that emerge from the interpretation of the Book.

Thus the *Šarī'a* is now only the first level of meaning relevant to the life of ordinary believers.

The tradition has become a living one, continually re-affirmed in the relationship of initiation linking the master (*Imām*) and the disciple.

The relationship to the source is reversed, in that the full meaning of the source will only be realized and actualized in the *telos* represented by the coming of the *Mahdī*. Time—not empirical time but the secret time which duplicates and crosses it—acquires power: it is the place in which meaning is formed and revealed. At the level of temporal experience it is the attitude of expectancy which characterizes Shī'te 'consciousness'.

Theologia subtilis . . .

For them, the world is not an accumulation of objects in space but a heterogeneous succession of independent actions. It is not spatial, but successive and temporal. (J. L. Borges, 'Tlōn Uqbar Orbis Tertius', *Fictions.*)

Instantaneism, which Louis Massignon regards as central to Islamic thinking about time, would seem rather to be a secondary elaboration characteristic of a 'regional' mode of thought: *kalām*. If it is taken as the unique temporal model, one must necessarily accept the vision which *kalām* had of itself and of its place in Arab learning: that of the sole depository of the meaning of the revelation.

The systematic nature of the doctrine of which it is part, as propounded in Maimonides' *Guide for the Perplexed*,[39] confirms that it is a secondary elaboration.

The First Proposition affirms 'physical' atomism. The atom has no material reality and is an *ens rationis*.[40] But an assemblage of atoms forms a body either by its own virtue or by conferring corporality on the atoms which are thus joined together. In the latter case, the coming together of two atoms affects their very nature because, for the first time, they become a body. This may appear to be paradoxical: how, indeed, can something which has no magnitude produce magnitude when added to itself?

The formation of bodily substances which are composed of particles without magnitude implies that the process of putting them together possesses some virtue of its own and that the terms brought into relationship with each other are altered by the fact of being so related.

The manner in which bodies are composed (not by mixing atoms but by juxtaposing them) and what brings the process of composition (or decomposition) about, namely movement, imply the existence of a void (Second Proposition).

Birth and destruction consist in the union and separation of the identical particles of which bodies are composed. But this does not imply that atoms are 'limited in number in the universe'. Indeed, there is no pre-existing material from which God could make different configurations and there is no finite quantity of elements changing only with regard to their law of composition. On the contrary, the elements and the law governing their organization come into being contemporaneously as the result of a single and indefinitely renewed act of creation.

This may appear to be incompatible with the affirmation of the absolute similarity of atoms and more especially with the affirmation which follows logically from this to the effect that birth and destruction consist only in the union and separation of the same atoms. But as the atoms are kept in existence by a perpetual act of creation (cf. Sixth Proposition), it is only through a certain misuse of language that one can say that the *same* atoms enter into the composition of different bodies. The problem is thus one of knowing what guarantees the identity of a given atom (or of a given body) throughout time.

This problem arises once one accepts that time is composed of independent moments (Third Proposition) which 'because of their short duration cannot be divided'. This proposition is put forward by Maimonides as a necessary consequence of the First Proposition, for if the *mutakallimūn* (theologians) accepted the continuity of time, they would be obliged to accept the continuity of space.[41] The negation of the space-continuum entails a similar negation in the related fields of local movement and time.

Time is thus made up of instants which are indivisibles. Hence there is a realism of the element in the doctrine of time the instant is not regarded, as Aristotle regards it, as an ideal limit between the past and the future but as a constituent element of time.

The fact that the space traversed by a moving body is composed of indivisibles means that all movements are equal, apparent differences in speed being due only to the different intervals of repose punctuating the movement.[42]

It is here that the Twelfth Proposition,[43] which manifestly has a critical role, comes into play. This proposition lies outside the deductive configuration made up of the other propositions; on the contrary, its function is regulatory and it intervenes at every level of the deduction whenever appearances are to be criticized. If space is composed of indivisibles, it also follows that the notion of irrational magnitudes, and consequently Book X of Euclid's *Elements*, must be rejected.[44]

The Fourth Proposition is necessary in order to complete the 'physical' theory and the theory of time. An 'accident' is a *ma'nā* attached to the body.[45] Accidents come as correlatives (life/death, knowledge/ignorance, movement/rest, etc.), one term of which is necessarily related to an atom. Thus not only does each atom comprise an accident; each accident is also referred to an atom.[46]

This 'actualism of the accident' entails the validity of the Fifth

Proposition, which comprises two closely related statements: (a) the accident resides in the atom (with the exception of quantity which is not an accident); (b) it is not the body as an assemblage of atoms which is the collective substratum of the accident but it is the atoms of which it is composed which form the distributive subject of this accident.[47]

In reply to the objection that the accident lies in the body as a whole (in that it disappears if the totality ceases to present itself as such—for example, if one cuts off parts of a living being, they cease to be living), the *mutakallimūn* propound the thesis of the creation of accidents. Not only does the objection err by making the substratum a totality that can be expressed in its accident; it also fails to grasp one of its essential characteristics—its non-durable character and direct dependence on the divine will.

Hence the Sixth Proposition: 'the accident does not last two periods or two instants'.[48] This proposition marks the absolute dependence of the accident on divine will. God could not create a substance without accident (this statement is related to the Fourth and Fifth Propositions). The asymmetry which in Aristotelianism makes the substance a first term with regard to the accidents is eliminated. And, according to Maimonides, this is in line with the underlying ratio of the non-durability of accidents, for a substance is no longer something which, by the workings of nature, requires a particular specific accident but is created in a single identical movement *with* an accident. When Ibn Suwār[49] wishes to demonstrate the inconclusiveness of the arguments used by the *mutakallimūn* in favour of the created nature of the world, he can only do so by a piece of semantic legerdemain which enables him to restore the principle of the prior existence of the substance in relation to its accidents, a principle that is specifically not recognized by *kalām*.[50]

This ontological equating (or near-equating) of substance and accident brings out the underlying intent of this corpus of propositions: the negation of *phusis*, of natural law, understood as a hierarchical and stable ordering of a cosmos. The non-existence of a '*nature* of things', that is to say of a law governing their development and their interconnexion with other phenomena, necessarily entails the instantaneity of a creative act[51] and its recurrence throughout time in accordance with a custom (*'āda*) which is a rule of repetition.

This Sixth Proposition is entailed by the Third Proposition. Indeed, the latter's affirmation that time is composed of independent moments already made it necessary to explain the continued existence of phenomena in time.

The Sixth Proposition makes a positive statement of this fact from the point of view of the thing which lasts: no process of self-positing, no internal spontaneity can explain why it should last; for this purpose outside intervention is necessary. Hence, a theory of continuous creation. The reason this continuous creation applies to accidents is that as an accident qualifies a substance totally, its continued existence or its destruction entails the continued existence or the destruction of the substance (cf. Fourth, Fifth and Sixth Propositions). Because there is no internal dynamism of the substance (atom), it is sufficient for the divine will to refrain from action in order for the substance to return to non-existence. As such refraining (*tark*)[52] is sufficient to cause a substance to cease to exist, it is not necessary for God to create non-existence.

This would, however, have to be postulated if it were claimed that a substance lasted more than an instant; for if it ceased to exist after having lasted a certain time, a sufficient reason would have to be given in order to explain this return to non-existence. As this reason can only be God, it must necessarily be admitted that God creates non-existence. Thus the theory of continuous creation is in accordance with the principle of Occam's Razor: it is not necessary for God to create 'an accident of destruction' in addition to the accident relating to the substance; it is sufficient that he refrain from creating.[53]

This continuous creation cannot be continuous reproduction, otherwise it would be necessary to postulate between one creation and another, a non-existent which would need to be created; as God does not positively produce non-existence, he cannot produce in beings a quality of non-existence only in order to efface it subsequently. One of the direct consequences of this Sixth Proposition is that the regularity of nature is not grounded ontologically but is merely the effect of a custom (*'āda*) established by God. If the movement of the hand (of the writer) can have no influence on the pen, this is by virtue of the principle according to which 'the accident does not extend beyond its substratum': movement, as an accident of the hand, will not be transmitted to the pen. If the accident extended beyond its substratum, this would imply a dynamism proper to nature without the continual intervention of the divine will. Thus the notion of a link with an antecedent which acts on a consequent has no place in such a system and, indeed, the system denies all causality.[54] 'In brief, one cannot say that a thing is the cause of any other thing.'[55] Custom refers to the co-existence of certain accidents. Thus, when someone writes, four accidents created by God co-exist: the desire to move the pen, the ability to move it, the movement of the hand and the movement of the pen. Furthermore, these four accidents are themselves continually recreated so that causality is not a kind of 'horizontal' link between phenomena but a 'vertical' link between the sole efficient cause (God) and all existing phenomena.[56]

The example of will and action cited here by Maimonides is precisely the example in connexion with which the theory of temporal discontinuity is advanced for the first time in *Maqālāt al-Islāmiyyīn.*[57] The matter at issue is whether the ability to carry out an action (*'isṭiṭa'a*) lasts more than one moment or not. The reply of the *mu'tazilites* is quite definite: it does last.[58] But the reply of abu-l-Qāsim al-Balẖī, although an exception, is no less significant:

The ability to perform actions does not last two (consecutive) moments; such duration in time is impossible. The action takes place in the following moment by virtue of an antecedent capacity which has passed to the state of non-existence. But as the action cannot occur if man is in a state of incapacity, God in the following instant creates a capacity and the action occurs by virtue of the antecedent capacity.

The moral capability of man is subject to the same discontinuity as time. It is divided up, so to say, into atoms of capacity. But, this being so, how can a decision be embodied in action? It is necessary that, at the moment in which the will passes to the state of non-existence, man is not in a state of incapacity. In fact, three situations are possible:

1. A state of 'inhibition', where there is a conflict between an antecedent capacity and a present incapacity. This does not necessarily result in a state of incapacity as is shown by the solution adopted by abū-l-Huḏayl.[59]
2. A neutral state in which the agent is neither is a state of inhibition nor in a state of capability. In this situation, it can readily be imagined that there would be a time-lag between the act of will occurring in the preceding instant and the action carried out in the present instant.
3. A 'positive' state of present capability.

As a solution, the first possibility is ruled out probably because the present state may be expected to prevail. The second involves a merely negative condition: there is, at the present time, no obstacle to the accomplishment of an action previously decided on. The third, which is that singled out by abū-l-Qāsim, represents a maximum for there is not only no obstacle but a presently existing positive capability.

Divine intervention chooses this optimal solution by producing in man a new act of will although the action is accomplished by virtue of the act of will which precedes it and not that which is created at the moment when the action is accomplished.

Thus, there is temporal discontinuity not only with regard to the 'perceptible' accomplishment of the action, but also with regard to the register in which the will of the moral agent is operative; the result of this is to introduce, between the two series, a hiatus in which the divine will intervenes, adding on each occasion to the 'noumenal' series (if the Kantian term is not too inappropriate) a kind of supplement. This 'noumenal' supplement, which is theological in character, is what makes it possible to equate the two series. Temporal discontinuity involving continuous divine intervention precludes all moral autonomy. This explains why the majority of the *mu'tazilites* did not adopt the solution propounded by abū-l-Qāsim.

This doctrine of continuous creation does not apply only to positive accidents but also to 'privations of capability' because rest, no less than movement, requires an act of creation which keeps it in existence. Privation is not a form of non-being, not even the relative non-being which Aristotle considered it to be. Rest, for example, is not the completion of the movement of a movable body which has reached or regained its place. On the contrary, it is an accident the origin and continued existence of which call for an explanation. Thus, rest and movement are subject to the same ontological laws. The same is true of life and death, ignorance and knowledge, darkness and light, etc. Their existence is subject to the same temporal discontinuity and they are therefore equally in need of divine intervention if they are to be perpetuated.

If privations of capability can lay claim to the same ontological status as positive accidents, the Eighth Proposition may be pronounced valid: 'Everywhere there is only substance and accident', which may be re-phrased as: There is positive everywhere. Or again, as God cannot create a substance without accident, Everywhere there are substances with (actualized) qualities. This homogenization of *onta* implies that no form exists which is characteristic of a being or, in other words, that there is no specific form. Actualism goes over into nominalism. This homogenization makes both heaven and earth subject to the same constituent laws. The world, in its totality, being subject to radical contingency, there is not, on the one hand, the sub-lunary world of becoming, subject to contingent processes, and, on the other, the supra-lunary world imbued with regular movement. This is the meaning of the Tenth Proposition, which is essentially both logical and theological: if there are no fixed natures, no *loci* of an a priori nature (in the Aristotelian sense) and no abiding specific forms, then there are no bounds to cosmic contingency and it is conceivable—and rationally admissible—that the earth might become a celestial sphere and, conversely, that the element of fire might cease to rise upwards and that the element of earth might no longer be drawn toward the centre or that there might exist 'an elephant the size of a gnat'.

According to Maimonides, all the preceding propositions lead up to this one. But, above all, it is the hypothesis of the 'parity of atoms' and of the ontological equality of accidents (Eighth and Ninth Propositions) which is at the basis of this proposition. And this hypothesis in turn is necessitated by the doctrine of time: for as the accident does not last two instants, it cannot serve as the 'subject' of other accidents.

This Tenth Proposition is the theological expression of the negation of *phusis* as the autonomous development of beings, of the regulated, hierarchical and stable cosmos.

The radical contingency of the world being thus affirmed, the Eleventh Proposition is required as a necessary complement: its purpose is to forestall recourse to the notion of infinity, which could otherwise, by affirming the eternity of the world, rebut its contingency. Just as atomism implies the negation of the infinite *in posse* (which would, indeed, imply that magnitude was infinitely divisible), so the contingency of the world implies the negation of 'the infinite by succession'. The doctrine of time postulates a world which is both discontinuous and finite.

This is in keeping with the distinction between *qadīm* and *muḥdaṯ* which is one of the essential structures of *kalām*.

The concept of *qadīm*, which may be translated as 'eternal', has not always had this technical meaning. We find this valuable piece of information in *al-'Imta' wa-l-mu'ānasa* by abū Ḥayyān at-Tawḥīdī.[60] The grammarian-theologian abū Sa'īd aṣ-Ṣīrāfī states that for the Arabs *qadīm* does not signify that which has no commencement in time but that whose commencement is lost in the night of time. Its significance is not referred to a theo-cosmology but to a professed ignorance as to the first appearance of a phenomenon.

The technical meaning which the term *qadīm* takes on in theology is a refinement of its generally accepted meaning: here the extent of the past is no longer left merely indeterminate and we arrive at what is indeed an infinity, with no *terminus a quo*; from an origin which, because it is unknown or has been forgotten, cannot be established, we pass to a total absence of origin (ontologically affirmed); in place of a failure of the imagination (*wahm*) there is a clear and distinct principle of the intellect (and of things); instead of an empirical affirmation (that the past stretches endlessly behind us) we have an a priori definition. *Qadīm* becomes that which, by definition and a priori, has no beginning in time. In other words, in place of the amorphous notion of *dahr*, time is given a shape.[61]

The division of being into *qadīm* and *muḥdaṯ* only becomes effective within a theory which has rid itself of an anthropomorphic God (*tašbīh*);[62] the *Rāfiḍites* believe, for example, that God has a body and is not an immutable being. He acquires certain attributes (powers, eyesight, hearing) only when he has created the things which make possible the exercise of these attributes.[63] For before God created beings, they were completely non-existent; and attributes cannot be employed with regard to a non-existent.

Such an approach implies that beings exist for God neither as potential creations nor as examplars present in the mind.

It was, perhaps, in order to get round such a situation that Naẓẓām put forward his theory of *kumūn*[64] (latency) according to which the aggregate of beings exist in some way simultaneously in the 'sight' of God although in temporal terms they make their appearance in succession. Thus it is not necessary for certain beings to appear in order for the divine attributes to be actualized.

This resolves a dilemma inherent in the notion of creation. How is the absolute act of creation expressed in terms of time? In the theory of

Naẓẓām, time is not, as might be expected, abolished, but constitutes the law governing the appearance of beings and things drawn from a kind of pre-established fund.

The concepts of latency and of manifestation (*ḏuhūr*), which are characteristic of Naẓẓām's preformationism, have a function which is analogous to that of the concepts—originally neo-Platonic but taken over by the spiritualists in sixteenth-century Europe[65]—of envelopment or encasement and development or unfolding: it is only in time that the development occurs of what is enveloped, and what is 'complicated' becomes 'explicated'.

As no such elegant solution was available to the *Rāfiḍites*, they refused to acknowledge the immutability of God: the divine attributes are actualized only in so far as they find an application (which assumes the appearance of beings to which they are applied); what is more, God can 'change his mind', he may have a *badā'a*. Thus the will of God has no definitive orientation or immutable direction and this applies equally to the creation of beings and to the institution of a religion (*šarī'a*). This, at least, is the significance which the *Rāfiḍites* give to the abrogation of certain verses of the Koran (*nasẖ*).[66]

These three propositions are diametrically opposed by three theses of the *mu'tazilites*:[67]

1. Divine knowledge is not actualized as and when the objects which are its subject make their appearance; it is not created in time but is immutable.
2. God has no *badā'a*.
3. He never abrogates a 'factual' statement (*ẖabar*) by another 'factual' statement for this would imply that one of the two was false (*kaḏib*). Abrogation can occur only in the case of *'amr* and *nahy*, that is to say in commands or prohibitions, a category of statements which cannot be said to be true or false.

Thus God is endowed with the essential predicate of immutability: God is regarded as an extra-mundane being who remains unchanged by his dealings with the world. The principle of this transformation lies in the shift from a *real* relation to a relation which is real in one direction and ideal in the other.[68] For the *Rāfiḍites*, God is not an unchanging being who remains unaffected by his creation; the relation between him and the world is not unidirectional but two-way. By this means the dilemma of the *relatio creationis* is solved in such a way as to satisfy the criteria of logic, since we no longer have here a relation without converse.

The cognitive relation itself corresponds to this requirement: it is neither realistic nor idealistic and may be termed 'parallelist'. This implies that the object of knowledge and the cognitive 'subject' are both equally present. This explains why God's knowledge of things cannot be 'all at once' but becomes actualized as and when the objects of knowledge make their appearance.

It is in its challenge to the idea of such a reciprocal relation between God and the world that we discern the significance of one of the fundamental oppositions found in *kalām*[69] and even in *falsafa*,[70] that between *qadīm* and *muḥdaṯ*[71] as well as all the related oppositions: necessary/contingent, necessary existence/possible existence, creation *ex nihilo*/eternity of the cosmos, etc.

God, being *qadīm*, is no longer regarded, as he is by the *Rāfiḍites*, as a being who somehow participates in the becoming of the world. As a result of this opposition God becomes more remote and is therefore accessible only through the medium of a negative theology.[72]

This denial of a reciprocal relation between God and the world explains even the terminological precautions of the *mutakallimūn*: the *Aš'arites* avoid referring to God as a cause and prefer to use the term *agent* (*fā'il*):[73]

> The philosophers, as thou knowest, refer to God as the *first cause*; but those who are known as *mutakallimūn* sedulously avoid this term and refer to God as the *agent*. They believe that it makes a great difference whether one says *cause* or *agent*: for, they say, if we said that he (God) is a cause, this would imply that the *effect* exists, from which we should be led to conclude that the world is eternal and necessarily co-exists with God; but if we say *agent*, it does not necessarily follow that the object of the action exists together with the agent, for the *agent* may exist prior to his *action*, and they even go so far as to say that the agent may only be so called on condition that he does exist prior to his action.

Maimonides, who reports this opinion, rejects this distinction for, in his opinion 'the object of the action necessarily exists when the agent is *in actu*, so there is no difference'.

The subtlety of a terminological distinction here conceals a fundamental point of disagreement: whether the world is eternal or was created *ex nihilo*.

In the argument as presented this 'doctrinal' dispute is placed on a theoretical level which disregards—that is to say which does not call into play—the distinction between actuality and potentiality that Maimonides on the contrary invokes when he states that the effect follows from the cause when the latter is *in actu*. This failure to recognize the distinction implies a rejection of the dynamic notion of time embodied in the shift from potentiality to actuality. Theoretically speaking, this is perfectly in keeping with the temporal atomism proclaimed by the *Aš'arites*. If no distinction is made between potentiality and actuality, this brings into play a pattern of causality according to which the pre-existence of a cause as compared with its effect is not merely logical (the cause being the logical prerequisite of the effect) but also chronological; hence the absolute nature of the divine will is enhanced because God, strictly speaking, is not a cause since the notion of cause implies a correlative effect.

What may appear to be a pointless verbal refinement stems, therefore, from the desire to assert the inapplicability of the category of symmetrical relationship to divine action, thus definitively precluding the eternity of the world. Time, if one may still—or already—apply this term to what separates void and creation, is the activity of the Divinity in which an absolute decision will set in motion the transformation of the world from non-existence to existence.

Maimonides emphasizes that physical and temporal atomism is assumed by the theologians in order to affirm the radical contingency of the world. Indeed, the 'parity of atoms' and the lack of any difference between them depending on the body into whose composition they enter, the ontological equality of accidents and the absence of stable form lead to the conclusion that the ratio for the various aggregations of atoms of which natural beings are formed is not to be found within such beings themselves. Hence the necessity for continuous divine action both in order to impose on atoms a certain configuration and in order to maintain this configuration in being for more than one instant of time. This is, therefore, not so much a physical theory as a *schema argumentationis* for the purpose of proving that the world was created. It is, in any case, in this light that it was regarded by so perceptive a thinker as Ibn Rušd.[74]

For this reason, one should not be too hasty in identifying what is known as the Muslim 'occasionalism' with Cartesian instantaneism and its corollary, continuous creation, nor even with the occasionalism of Malebranche for, in the case of the latter two, the metaphysical thesis is the foundation of a physical doctrine and is conceived with a view to physics.[75] The separation *and* conjunction of different fields in traditional Western philosophy, a separation and conjunction which have been elaborated and thematized as such (the entire purpose of the *Meditationes* lies in providing a basis for physics, cf. also the epistolary preface to the *Principia Philosophiae*), make such an identification hazardous, or, at least, indicate that we should abide by the most formal frameworks not with regard to method but with regard to the claimed results.[76] The two distinctive marks of time in *kalām* are that it is discontinuous and finite; these two characteristics will be reverted to on a different level, transformed or refined, in mysticism and in *falsafa*.

The finite nature of time demonstrated *ab absurdo*

We shall here take Kindī as the fountain-head of *falsafa*. It is perhaps significant that his characteristic and favourite mode of argument is the demonstration *ab absurdo* as though *reductio ad absurdum* was the most suitable weapon for expanding the limits of rational thought.

On the basis of the following axioms ('common notions') Kindī demonstrates that there can be no such thing as a 'body' (*jirm*) infinite in actuality:

1. All bodies none of which is greater than the other are equal.
2. Equal bodies enclose equal magnitudes within their extremities, both in actuality and in potentiality.
3. Anything which comprises an end, a termination (*nihāya*), cannot be infinite.
4. For all equal bodies, if a body is added to any of them, then the body to which this addition has been made becomes greater than the others and great than it was previously.
5. If two bodies of finite magnitude are put together the resulting body is finite magnitude. This is true of any type of magnitude and of anything involving magnitude.
6. The smaller of two beings of the same kind affords a measure of the greater, or of part of the greater.

All these common notions concern equality and infinity: (1) propounds the condition for equality; (2) appears to link the relation of equality to finite magnitude; (3) propounds a negative 'definition' of infinite magnitude; (4) is equivalent to the axiom: the whole is greater than the part; (5) will be used in the demonstration of the finite character of future time: this demonstration will only appear conclusive in so far as Kindī does not envisage the possibility of portions of time accruing indefinitely to past time (supposedly finite) but isolates a particular point in this 'chain', that at which *a* single portion of time accrues to *a* time of finite magnitude; (6) affirms that there is always a relation between two comparable magnitudes.

I

The demonstration that there can be no such thing as a body infinite in actuality, which will serve as a paradigm for demonstrating that there can be no such thing as a time infinite in actuality, takes the form of a demonstration *ab absurdo*. Let us suppose that such a body exists and let us take away from it a finite quantity. This leaves us with a two-fold possibility: (a) either what remains after the removal is a finite quantity; (b) or it is an infinite quantity.

Let us examine the first hypothesis. Let us suppose that what remains is a finite quantity and, by reversing the first operation, let us add to it the finite quantity which was removed. What is obtained, by virtue of (5), is a finite quantity because if two finite quantities are added together their sum forms a finite quantity, which is not in accordance with the initial premise (which assumed an infinite body).

Let us now suppose that what remains is an infinite quantity. If the finite quantity which had been removed is added to it, it is: (a) either equal to itself; (b) or greater. In the first case, this would be tantamount to adding one body to another without increasing the latter body, and the whole would be equal to the body as it was before a finite quantity had been added to it, which violates the axiom that the whole is greater than the part.[77] Kindī proceeds here as if the operation of adding was not simply the opposite of the previous operation of taking away: he isolates the former operation from the latter. In the second case, we obtain an infinite quantity which is greater than another infinite quantity. Now, by virtue of (6), the smaller (infinite) quantity affords a measure of the greater or of part of the greater. Therefore the smaller infinite quantity is equal to part of the greater. But equal quantities comprise extremities which limit them ((2) and (3)). The smaller infinite quantity (being equal to part of the greater) is therefore finite, which is contradictory. 'It is impossible for one of the infinite quantities to be greater than the other.' Not only in the notion of equality inapplicable to infinite quantities but also, if the notion of equality is applied to quantities, they must necessarily be finite.

The method adopted has thus been one of demonstrating an absurdity and a contradiction by means of a purely operational procedure involving no statement concerning the nature of the terms being manipulated. The means whereby this is done is the Law of Excluded Middle.

This demonstration, Kindī argues, may be applied to any quantity whatever its nature. Now time belongs to the category of quantity and is therefore finite and has a beginning. Time, as a 'predicate' of the body (like place and movement), is also finite by virtue of the finite nature of the body.

In this way mathematical finitism is transposed into cosmological finitism: the world cannot be infinite either in its dimensions or in its duration.

Time, the argument goes, as a means of measuring movement,[78] supposes movement without which it would have no reality. Movement, in its turn, supposes a body (a substance) which moves. The movement and the body are indissociable from each other. They are contemporaneous. This proposition holds good for the 'body' of the world.

Let us suppose that it was originally at rest and then moved (movement being contained within it in potentiality). There are only two possibilities: (a) either it is a being produced *ex nihilo*; (b) or it is an eternal being.

Let us examine the first hypothesis. If the world was produced *ex nihilo* its coming into existence would be a generation and a generation is one kind of movement. If therefore, the world does not precede (exist before) generation—because its coming into existence *is* generation—it cannot precede movement; this is incompatible with the initial premise whereby it was assumed to be at rest. One cannot therefore say that it is without movement because it cannot exist without movement.

The second hypothesis assumes that the body of the world was *eternally* at rest and then began to move. We then accept that an eternal being can change (because it passes from a state of rest in actuality to movement in actuality). But an eternal being, by definition, does not change. This brings us to a contradiction because we accept that a being can both change and remain unchanged.

In the two mutually contradictory hypotheses, we reach a contradiction as soon as body and movement are not linked. This leads to the conclusion that the body of the world never precedes movement. Thus the body implies movement and vice versa. We have already shown that time implies movement. Consequently, time never precedes the body. 'There is no body without duration, for its essence (i.e. that which makes it what it is) is expressed through duration.' The duration of a body is measured by the movement of this body and therefore the body never precedes time. Thus, body, movement and time never precede each other.

This reciprocal entailment makes it possible to extend to time the demonstration of the finite nature of the body. Time is finite in actuality. The world cannot therefore be eternal. Since it does not precede time it can have no infinite existence.[79]

In this connexion it is sufficient to recall that Aristotle too had applied to the duration of the world only one kind of infinity (the infinity which determines the conditions of an Archimedean line) while insisting that infinity was in other respects a notion inapplicable to 'cosmology' since the world is a spherical, finite body.

From one point of view, by affirming the finite nature of time, Kindī takes to its limit the Aristotelian view that the notion of infinity is inapplicable to cosmology. He thus places on the same level considerations with regard to the dimensions of the world and those relating to its duration.[80]

However, the considerations which guided the work of Kindī were of a theological rather than an epistemological nature: his main preoccupation was to reconcile philosophy and the Muslim revelation according to which the world was created.

II

The second demonstration of the non-eternity of the world is valid on condition that its initial premise—that the process of composition and assembling characteristic of the body constitutes a change—is verified.

1. The body is composite, and its composition has two aspects:
   (a) A logical aspect: the body is a substance having length, breadth and depth. It is therefore constituted of a genus: substantiality, and of a specific difference: length, breadth and depth.
   (b) A physical aspect: the body is composed of matter and form.
2. Composition supposes movement: composition is a change in relation to the state of non-composition and is therefore a movement.
3. Thus, body and movement never exist prior to one another: indeed, the body, being composite, supposes movement.
4. As movement is change, and as change is the means of measuring the duration of what changes, movement affords a measure of the duration of what changes.[81]
5. Time is an interval (*mudda*) measured by movement.[82] And all bodies have a duration (*mudda*).
6. As the body does not precede movement (cf. 3) it does not precede the duration which is measured by movement.

Thus body, movement and time co-exist. If, therefore, time is finite in actuality, the existence of the world (of the body) is necessarily finite and is therefore not eternal.[83]

As in the preceding demonstration, the central thread of the argument is the reciprocal entailment of body, movement and time. But this second demonstration is direct although hypothetical.

III

The third demonstration of the non-eternity of the world is based on the denial that the past and the future can be infinite in actuality.

To demonstrate that the past cannot be infinite in actuality, let us suppose that by going back up the time series of the past one never arrives at a first term (or at a final division). It follows that one will also never be able to redescend the time series down to a particular given moment. For, if this were not so, the time which stretches from infinity down to the present moment would be equal to the time extending from the present moment towards infinity.

If, then, the time which extends from infinity to the present is known (or determined), the time extending from the present moment toward infinity is also known or determined. This would mean that an infinite quantity is in fact finite, which is a contradiction. Past time is therefore finite in actuality.

This argument hinges on the property of equality which we have already encountered: if magnitudes may be said to be equal, then they are finite.

Kindī postulates an adversary who is prepared to accept not only an infinity in actuality of past time but also a particular moment at which one is supposed to be or at which it is supposed one might be, and who is thus driven to acknowledge an equality of two infinities.[84] The postulate underlying Kindī's demonstration is the same as Aristotle's: 'An ordered infinite whole is a whole which has no end.'[85]

Kindī offers a further demonstration that the past cannot be infinite in actuality: if we reach the present moment only after having previously reached the moment which precedes it and that only after reaching the moment preceding that one and so on *ad infinitum*, as an infinite distance cannot be traversed we should never arrive at the present moment. Now, we know from experience that the present moment exists. Therefore we did not have to traverse an infinite distance in order to reach it.[86]

In reality, what Kindī 'demonstrates' is not so much the necessity of an absolute beginning to time as the absurdity—in his view—of supposing that a—finite—interval of time is composed of an actualized infinity of moments.

The demonstration that the future cannot be infinite assumes that the finitude of past time has been demonstrated. This is the first premise. The second is as follows: the different parts of time succeed each other. It follows that whenever a new part of time is added to a determinate finite time (as in the case of past time) the total so obtained can only be finite, by virtue of the fifth of the axioms ('common notions') listed at the beginning of this section.

Furthermore, time is a continuous quantity which means that it is divisible by means of a limit which is common to the past and the future. What thus divides it is the instant (the now), which marks the end of the past and the beginning of the future.

Now, any determinate portion of time has two extremities: a *terminus a quo* and a *terminus ad quem.*

If, therefore, a portion of time is linked to another portion by an extremity which they share in common, its other extremity is determined and known.

This argument is aimed at the Aristotelian idea taken over by Ibn Rušd in *Tahāfut at-tahāfut*,[87] according to which, as time cannot be conceived without an intermediate term (the instant) and as the latter, in its turn, implies an anterior and a posterior (and therefore a period of time), it follows that a temporal limit cannot be conceived without time itself also being conceived.

Once this idea has been dismissed (or, more exactly, turned upside down—the instant in Aristotle affording the proof that time always exists; in Kindī the instant is the common limit, but of two portions of time which are limited although the limit, in this case, is not an intermediate term), Kindī affirms that whatever number of determinate portions of time are added to a portion of time which is itself determinate (in this instance past finite time), the total must necessarily be determinate.

Consequently, future time is not infinite in actuality since
merely of a collection of determinate portions of time added to
which is itself determinate.[88]

This demonstration also hinges on the fifth axiom or 'commo
In formulating this 'common notion', Kindī avoided making
nouncement on the outcome of the indefinitely repeated process
one finite magnitude to another. But, in connexion with the suc
portions of time, if the demonstration is to be conclusive it is ess
Kindī should not accept the proposition stated in *Physics VIII* 2
by continually adding to the finite, every finite will be exceeded

Be that as it may, it is thus philosophically demonstrated, in
consistent with the teaching of Islam, that time (and, consequ
world) has a beginning and an end.

Kindī's affirmation of total cosmological finitude as re
dimensions of the world and its duration was later to be used
in his polemic against the *falāsifa*, while Ibn Rušd tried, on the
to reduce to absurdity this linking of space and time.

Indeed, the mainspring of Ġazālī's argument was to be his
reduction of time to space, resulting in attaching to time all the
of space, in particular finitude. Ibn Rušd was to see in this ide
of space and time a *locus communis* of sophistry,[89] and, in the th
creation of the world *ex nihilo*, a transgression of the principle o
reason.[90]

Thus *falsafa*, which began with the closely reasoned dialectic of
Kindī, was to continue its battles with sharper and sharper weapons, a process in which obedience to the rules governing philosophical discourse was to involve remarkable shifts of position: Ġazālī the theologian wishing to disseminate the truth however demonstrative it might be, and Ibn Rušd the *faylasūf* developing a theory of 'ideology' which was noteworthy for its literalistic tendencies.

The time of the presence or time in mystical experience

Sufism does not disregard the reference-event which defines *waqt* for the theologians,[91] nor the instant as the limit between the past and the future, but sees the reality of time in the moment dominated by an affective state, coloured by a *Stimmung*. The Sufi, the 'son of his *waqt*', is in a constant state of receptivity vis-à-vis his *waqt*, 'absorbed' by the voice within him, responding with full attentiveness to the call which he hears. The Sufi belongs ultimately to his time and he essentially relates to it.

This 'time' must not be duplicated by any memory which would unfold the past under the gaze of present time; this would be a betrayal of his own time which is being fulfilled at the present moment.

Similarly, the decision to renounce the world must not be delayed but taken at the present moment, without delay, confounding as it were, the swiftness with which truth descends on 'those who remain constantly in a state of hope'. By taking this decision and abiding constantly by it, the Sufi is always able to receive the truth in its disruptive force. The tension which is maintained in the present moment is the correlate of a truth which always comes as a sudden revelation. Inattentiveness, on the other hand, the postponement of the decision to renounce the world, obliviousness of the need to face the present, leaves the soul bewildered at the approach of truth.

The manner in which time is accepted may be propitious to the voice of truth or, on the contrary, may make it for ever inaccessible. The dimension of the present is decisive.

But it is less a matter of 'delimiting the present', to use the expression of Marcus Aurelius, than of being 'delimited' in it, if, that is, one may legitimately apply such a term as 'delimitation' to an experience involving a loss of one's self-identity in the encounter (*muṣādafa*)—the unexpected but ever awaited encounter—with truth (*al-ḥaqq*), involving also the abandonment of one's free will (*'lẖtiyār*) in order to enable truth to take charge of and guide the self.

To be 'in the power of *waqt*' is to abandon oneself (*'istislām*) to an apparent absence which becomes a presence (*ġayb*).

The *waqt* is not an amorphous flood into which one may 'plunge' at will; it is more like a blade and the manner in which truth 'makes it pass' and recompenses it is its cutting edge.

This comparison with a sharp blade indicates not merely the intrinsic character of *waqt* which is to be distinguished from the innocuous and innocent flow of a humdrum grey existence 'in which all cows are black'; the analogy is also intended to show that the manner in which we conduct ourselves with regard to *waqt* determines the way in which it affects us. Just as 'the blade has a smooth surface but cuts if we strike against it', so 'he who submits to the imperative of *waqt* is safe but he who opposes it is broken and dies'.

The confident abandonment of self to this time of the presence, the resolute acceptance of the judgement it brings, are an essential condition of salvation, a condition which is fulfilled in the first place with regard to *waqt* itself because it is also potentially the gulf through which our being escapes. This abandonment and acceptance are not an attitude of pure passivity but, on the contrary, an active attitude with regard to *waqt* which is an essential part of the experience of the presence.[92]

The Sufi is 'master of time', as Qušayrī expressed it,[93] when his movement from state (*ḥāl*) to state[94] puts him in a position in which the 'assault' of the presence does not throw him off balance.

By this process of 'sublation' (suppression, conservation, transformation of states), the Sufi reaches the genuine *tawḥīd* 'which consists', says Junayd, 'in the separation of the Eternal from that which has originated in time'.[95] *Tawḥīd*, the affirmation of the divine unity and uniqueness, is an affirmation of the difference between the eternal and the intra-temporal. The key to the structure of the mystical experience lies in this equivalence between unity and eternity. The mainspring of this experience is to efface in *fanā'/baqā'* the imprint of temporality by which the creature is afflicted.

It may seem paradoxical that while the essence of *tawḥīd* is said to lie in the separation of the eternal and the temporal, its affirmation in practice is an endeavour to obliterate this difference. But this tension is the life of Sufism, a tension which is always lived in the present.

This paradox leads ultimately to the dual sense of union and separation voiced by Junayd:[96]

So in a manner we
United are, and One;
Yet otherwise disunion
Is our estate eternally.

Transcended, this paradox gives us Ḥallāj asserting, in the claim that was to bring him to execution, 'I am the truth'.

As we have emphasized at the beginning of this essay, time is for us today the aprioristic structure of the experiential; history forms the supporting fabric of our knowledge. Since Hegel, and in spite of different re-statements of the concept of history, time remains the place in which systems are unfolded and transformed. There is an essential connexion between *time* and *system*. Time has only recently become freed from the dead hand of the system. Discontinuity, chance, event, series are the strategical principles of what M. Foucault calls a 'materialism of the incorporeal'[97] and, we might add, those of an empiricist approach to history (in the sense that there is no necessary connexion between two events.)

It follows that time cannot be represented by a line, nor by a circle, nor even by a multiplicity of interlocking lines, but only by a chaotic series.

In Muslim thought, time was not the form of inner organization of the system, and even if it did not disregard the concept of history—as witness Miskawayh and Ibn H̱aldūn—its significance in the whole structure of knowledge was quite different from what it was in Europe in the eighteenth and nineteenth centuries. As M. A. Sinaceur points out,[98] the interest of the idea that events constantly reproduce themselves in identical fashion is not so much that it emphasizes the regularity of history as that it undermines the role of *'isnād*, in other words the chains of transmission; to put it another way, it draws a line between that which is stated and he who states it.

In Muslim thought, time takes many different forms: *dahr*, whose involutions provide a stay and shelter for all things and beings; the cumulative time of knowledge based on ancient premises; the pure Event of the *mīṯāq* and its successive renewals; the discontinuous time of *kalām* which is, to use an expression of M. Burgelin,[99] 'the patience of God'; the continuous and finite time of *falsafa*; and the time of the presence in mystical experience.

The intention of the present writer has merely been to suggest that this diversity of meanings cannot be reduced to one and that it is only intelligible if the necessary allowances are made for the changes in emphasis and approach which occur in the treatment given to the concept of time in different disciplines.

Paris, September 1973

NOTES

1. Properly speaking, the notion of history, with its associated themes of scarcity, work, finitude, the end of history, etc., first appeared in the nineteenth century, the key figures being Malthus, Ricardo, Hegel and Marx; cf. the analyses of Michel Foucault in *Les Mots et les Choses*, p. 265–75, Paris, 1966.
2. Martin Heidegger, *Sein und Zeit*, cited in J. Derrida, *L'Endurance de la Pensée*, p. 224, Plon, 1968.
3. Kindī, *Épître sur la Philosophie Première*, abū Rīdah edition, p. 122; Fou'ād al-Ahwānī edition, p. 99–100.
4. Ġazālī, *Maqāsid al-falāsifa*, Suleimān Duniā edition, p. 167–8, Cairo, Dār-al-Ma'ārif, 1961.
5. Ibn Rušd, *Tahāfut at-tahāfut*, Suleimān Duniā edition, p. 153, Cairo, Dār-al-Ma'ārif.
6. Ibn Sīnā, *'Uyūn-al-ḥikma*, Badawī edition, p. 26 et seq.
7. *ān* is usually translated as 'instant' which is correct; however, it would be more accurate, and would give fuller account of its significance in Arabic, to translate it as 'now'. This was, in any case, its generally accepted meaning before it became part of the technical vocabulary of philosophy. Its area of meaning overlaps a number of other words, in particular *awān* (which is the time appointed for something, the time when existence acquires a moral, juridical or religious sanction).
8. Ġazālī, *Maqāsid al-falāsifa*, op cit., p. 167–8.
9. cf. Aristotle, *Physics*, Book IV.
10. This distinction is made in *Tahāfut at-tahāfut*, op cit., p. 154. The identification of the point and the now is presented by Ibn Rušd as being the source of the illusion which vitiates Ġazālī's reasoning.
11. On this point see the analyses by Aubenque in *Le Problème de l'Être chez Aristote*, Paris, Presses Universitaires de France, 1966.
12. This applies primarily to Book IV of the *Physics*, but also to Book VI.

13. It is true that we have in *Maqāsid al-falāsifa* a kind of handbook of *falsafa* as an autonomous field both inside and outside Arab culture; it is also true that Gazālī's purpose in writing this book, namely to make a polemical refutation of *falsafa*, might have led him to retain only the skeleton of the positions adopted by the *falāsifa*.
14. cf. J. T. Desanti, 'Une Crise de Développement Exemplaire: la Découverte des Nombres Irrationnels', in J. Piaget (ed.), *Logique et Connaissance Scientifique*, Paris, Gallimard, 1967. (Pléiade Series.) See also M. Clavelin, *La Philosophie Naturelle de Galilée*, p. 50 et seq., Paris, Librarie Armand Colin, 1968.
15. *Manāhij al-'adilla, Tahāfut at-tahāfut, Faṣl al maqāl.*
16. *Manāhij al-'adilla*, Mahmūd Qāsim edition, p. 138–9, Cairo, 1964.
17. Although Hellenism also explored the possibility of representing time by a line. cf. V. Goldschmidt, *Le Système Stoïcien et l'Idée de Temps*, p. 42–54, Paris, Vrin, 1969, who quotes:
    (a) the *Parmenides*, which examines this possibility in the second hypothesis;
    (b) a text by Syrianus in which he recognizes both the circle and the line as equally possible images of time;
    (c) 'the problem of *whether happiness increases with time*' which 'was commonly discussed in the schools before Plotinus devoted a treatise to the subject' (*Enneads* I, 5).

    With regard to Muslim thought, it may simply be pointed out that Ibn Sīnā bases his rejection of the thesis of the 'return of the same' (*al-'awd al-mumāṯil*) on our inability to predicate exact relations (in the mathematical sense, *manṭiqiyya*) between the revolutions of heavenly bodies, which alone would enable us to affirm with certainty that there is a recurrence of a given configuration (*tašakkul*) of the cosmos.

    In order to be able to affirm the recurrence of a given cosmic configuration it would be necessary to predicate this 'sameness' not only with regard to the natural species but also with regard to the actual individuals. However, astronomical observations enable us to arrive at results which are no more than probable. cf: Ibn Sīnā, *Sifā*, p. 195 et seq., Gener. et Corr., Section XV, Cairo, Mahmūd Qāsim edition, 1969.
18. cf. O. Cullmann, *Christ et le Temps*, Neuchâtel and Paris, 1947.
19. Louis Massignon, *Opera Minora.*
20. Kindī, *Épître sur la Philosophie Première*, abū Rīdah edition, p. 102 et seq.; Ahwānī edition, p. 80. This idea of the accumulation of knowledge seems to have been 'a commonplace among the Sophists and in the works of the Hippocratic physicians. cf. Aubenque: *Le Problème de l'Être chez Aristote*, p. 73 and 75.
21. ibid. Ahwānī edition, p. 81.
22. Although Aristotle accepts the notion of the progress of knowledge, he modifies it by affirming the return of the same opinions 'an infinite number of times' 'among men'; cf. *Meteorology* I, 3, 339^b and *De Coelo* I, 3, 270^b 19, which is a parallel, in the field of knowledge, of the cosmological doctrine of the eternal return. Kindī's affirmation of irreversible progress, at least in theory, is perhaps not unrelated to his theory of the finitude of time and the creation of the world *ex nihilo*.
23. Imru'u-l-Qays: *Dīwān*, Beirut edition, p. 31, 1972.
24. ibid. p. 99.
25. Sībawayh, *Al-Kitāb*, Hārūn edition, p. 37, Cairo, Dār-al-qalam, 1956. Sībawayh contrasts place (space) and time: place is analogous to 'persons', for: (a) places like persons may bear proper names: Makka, 'Omān, etc.; (b) places have 'physiognomies' which distinguish them from each other: mountain, river, sea, etc.; (c) places have a palpable existence—in short, places may be individualized and therefore named.

On the other hand, time (*dahr*) cannot be named, cannot be individualized, has no 'physical features' and cannot even be grasped. It passes continuously and thus has an affinity with the verb.

26. cf. the meaning of the celebrated phrase of Imru'u-l-Qays, who on learning the death of his father and postponing the decision to seek vengeance said: 'Today I have drained too deep the cup of feeling, tomorrow will come the time of decision.' This phrase which has become proverbial signifies a respect for the order of succession, the expectation that something which has found even a limited degree of realization in the order of time will be fulfilled in its entirety.
27. *Dīwān*, p. 99.
28. ibid. p. 31.
29. Although one should not attempt to place a systematic philosophy and a *Weltanschauung* on the same footing, *dahr* may be approximated to the *aiōn* of the Stoics before human action carves out of it a present. But in the case of *dahr* this will not occur.
30. cf. C. Mugler, *Deux Thèmes de la Cosmologie Grecque: Devenir Cyclique et Pluralité des Mondes*, p. 22 et seq. and p. 89 et seq., Paris, Librairie C. Klincksiek, 1953.
31. Ibn Sīnā, *'Uyūn-al-ḥikma. Dahr* is the *nisba*, the relation—in the mathematical sense—between *ṭābita*, things which are enduring, considered from this angle, and time. These enduring things are co-extensive with time (*ma'a az-zamān*) without being in time. *Sarmad* indicates the *nisba*, the mathematical relation, between what is not in time, considered from this angle, and what is in time. *Dahr*, considered in itself, is *sarmad*. We have here a refinement of the notion of eternity which is split into two levels.
32. cf. abū Bakr Rāzī, *Épîtres Philosophiques*, p. 304 et seq., Beirut, 1973.
33. Louis Massignon, *Parole Donnée*, Paris, Union Générale des Éditeurs, 1970 (10–18 series).
34. cf. Gardet and Anawati, *Introduction à la Théologie Musulmane*, p. 231, note 4, Paris, Vrin, 1948; cf. also H. Corbin, *Histoire de la Philosophie Musulmane*, p. 16 et seq. Paris, Gallimard (Idées series).
35. Islam itself claims to be a *return*, the reversion to an original source; cf. Sīrāj ad-Dīn, 'The Islamic and Christian Conception of the March of Time', *Islamic Quarterly*, I, 1954.
36. We shall return elsewhere to a study of this very specific form of polemics.
37. I. Goldziher, 'Muhammedanische Studien', French translation in *Bulletin des Études Arabes*, November-December 1942, Algiers.
38. This is true in theory even if, in practice, a distinction will be drawn between reprehensible and commendable innovation.
39. *Le Guide des Égarés* (trans. S. Munk), Vol. I, p. 375 et seq., sets out twelve fundamental propositions of *kalām*. English translation M. Friedländer, *The Guide for the Perplexed*, reissued 1925. A commentary on this passage will be found in McDonald, 'Continuous Re-creation and Atomic Time in Muslim Scholastic Theology', *Isis*, IX, 1927. See also al-Aš'ārī, *Māqālat al-Islāmiyyīn*, which is a catalogue of 'disputed questions' in Muslim theology.
40. S. Munk likens this thesis to the metaphysical atomism of Leibniz, cf. *Mélanges de Philosophie Juive et Arabe*, p. 322 note, Paris, 1859. See *Maqālāt al-Islāmiyyīn*, 'Abd al-Ḥamīd edition, Cairo, Vol. I, 1950; Vol. II, 1954, p. 4 and 8. In subsequent notes this publication will be referred to as *M.I.*
41. The link between the continuity of time and that of movement is demonstrated in *Physics* VI, 2, 232^{b}20 and 233^{a}10.
42. cf. *M.I.*, Vol. II, p. 18, 'Theory of abū-l-Huḏayl-al-'Allaf'. cf. also Nazzām's theory of the leap (*ṭafra*), which is linked with the thesis of the infinite divisibility

of bodies. An attempt to explain this appears in A. Nader: *Le Système Philosophique des Mu'tazila*, p. 182–7, Beirut, Les Lettres Orientales, 1956.

43. cf. *Guide*, p. 382 et seq.
44. ibid. p. 382–4.
45. cf. *M.I.*, Vol. II, p. 53.
46. See, however, abū-l-Hudayl, *M.I.*, Vol. II, p. 53 (question 39) and p. 44 (question 28).
47. In reality, it was a disputed question whether the atom on its own could be a bearer of accidents or whether accidents could appear in atoms only if the atoms had formed themselves into a body (cf. *M.I.*, Vol. II, p. 10–11). See also A. Nader, *Le Système Philosophique des Mu'tazila*, p. 153–5.
48. cf. *M.I.*, Vol. II, p. 44–6 (questions 28–31). Abū-l-Hudayl distinguishes between accidents which have no duration, such as movement, and accidents which have duration, such as certain states of rest. He expounds an ingenious theory according to which, at a certain moment, the movements of the inhabitants of heaven and hell are frozen in a lasting immobility; the bliss of the former and the torments of the latter are, so to say, concentrated and multiply *ad infinitum*. Duration is thus 'arrested' and the judgement pronounced by God becomes definitive. cf. Baġdādī, *Al farq bayn-al-firaq*, p. 102–3, Beirut, 1973.
49. cf. the text translated by B. Lewin, 'La Notion de *Muhdaṯ* dans le *Kalām* et la Philosophie', *Mélanges, Nyberg*, Uppsala, 1954.
50. cf. the thesis of Ḍirār Ibn 'Amr, 'Substance is an Architecture of Accidents' (see *M.I.*, Vol. II, p. 6–7 and 25).
51. cf. *M.I.*, Vol. II, p. 48. 'Abū-l-Hudayl says that the creation of a thing—which is its coming into existence *ex nihilo*—(...) is the will of God being brought to bear on this thing and his word telling it: "Be". The act of creation is contemporaneous with the thing created in so far as it is created.'

    It is instructive to compare this text with a text by Kant in his *General Physiogony and Theory of the Heavens:* 'Creation is not the work of a moment. ... The infinitude of times to come, which have an inexhaustible source in eternity, will animate all the spaces in which God is present, endowing them gradually with the regular order willed by the excellence of His plan. ... The work of creation is proportional to the time which is devoted to it.' (op. cit., Part 2, Chapter 7).

    The creative act may be likened to technical activity being carried out over a lengthy period, or even over an infinite period (Kant). This model requires an infinite universe which is progressively filled by worlds, such as is postulated in the *General Physiogony and Theory of the Heavens*, but abū-l-Hudayl conceives of the creative act as being analogous to a *performative statement*, which is in itself the accomplishment or an event (the change from non-being to being).

    Naẓẓām's theory of *kumûn* does not entirely correspond to this actualism. Rather, it postulates a potentiality which is progressively actualized (cf. Baġdādī, *Al farq bayn-al-firaq*, p. 127–8). It draws a distinction between an indivisible total act of creation, and a progressive appearance (*ḍuhūr*) of (already created) beings in time. It might be said that the act of creation constitutes the origin of beings whereas their order of appearance constitutes their beginning.
52. The question of what type of reality should be attributed to *tark*, was a widely discussed issue, cf. *M.I.*, p. 59 (questions 45 and 46), p. 60 (question 48), p. 61–4 (questions 49–58).
53. However, certain *mutakallimūn* think that if God wished to destroy the world, he would have to create a sort of unattached (substratum-less) accident, namely, an accident of destruction having the effect of 'neutralizing the existence of the world': cf. *Guide*, Vol. I, p. 391; cf. also the entire discussion on the reality of

fanā' in *M.I.*, Vol. II, p. 50–2. The concept of nothingness (and other related concepts) present certain difficulties to which we intend to revert elsewhere.

54. A distinction must be drawn between the *aš'arites* who regard such a statement as valid and the *mu'tazilites* who, on the contrary, accept the notion of cause. cf. *M.I.*, Vol. II, p. 69–71 (question 66). Mention has even been made in this connexion of 'rigorous determinism' (*déterminisme rigoureux*), cf. A. Nader, *Le Système Philosophique des Mu'tazila*, p. 190–207.
55. cf. *Guide*, p. 394.
56. Although abū-l-Hudayl affirms the existence of causality (in particular through *tawallud*, the 'engendering or bringing forth of acts', cf. *M.I.*, Vol. II, p. 79–83 and 84 (question 77)), it nevertheless remains true that any duration is the direct effect of an order addressed by God to the thing which has duration. There is thus this 'vertical' relationship between God and things, recurring throughout the 'horizontal' series of causes.
57. Although, in fact, this is briefly referred to in connexion with the divine word. See *M.I.*, Vol. I, p. 237.
58. This is the reply, for instance, of abū-l-Hudayl, Hišām, 'Abbād and al-'Iskāfī. Furthermore, this is in conformity with what Maimonides says (in connexion with the *mu'tazilities*) and with the thesis of free will which they propound.
59. *M.I.*, Vol. I, p. 276.
60. Aḥmad Amīn and Aḥmad al-Zayn edition, Beirut.
61. It would be instructive to make a systematic study of theological and philosophical notions in current usage, thus revealing the semantic fields of the terms involved and identifying their enduring core of meaning and departures from it.
62. 'Abd-al-qāhir al-Jurjānī sees in anthropomorphism a careless use of metaphors which is itself due to an inadequate systematization of rhetoric and of rhetorical figures of speech. cf. *'Asrār-al-Balāġa*, H. Ritter edition, p. 47, Istanbul, 1954.
63. cf. *M.I.*, Vol. I, p. 107–8.
64. cf. Ḥayyāṭ, *Kitāb al-'Intiṣār*, p. 51–2 and 132–3, A. Nader (ed. and trans.)
65. A. Koyré, *Mystiques, Spirituels, Alchimistes Allemands du XVIe Siècle*, Paris, Gallimard, 1971 (Idées series). The same concepts, informed by subtle mathematics, are found in Leibniz. cf. M. Serres, *Le Système de Leibniz et ses Modèles Mathématiques*, Paris, Presses Universitaires de France, 1968.
66. *M.I.*, Vol. I, p. 109.
67. *M.I.*, Vol. I, p. 256.
68. According to J. Vuillemin, the relation that exists between God and the world is of this kind, both in the theology of Aristotle and in Christian theology. cf. 4th essay: Théorie des Relations Mixtes', *De la Logique à la Théologie: Cinq Études sur Aristote*, p. 147–63, Paris, Flammarion, 1967. For an opposite point of view cf. Jacques Brunschwig, 'Le Dieu d'Aristote au Tribunal de la Logique', *L'Age de la Science*, Vol. III, No. 4, October–December 1970.
69. cf. *M.I.*, passim. Bāqillānī, *Tamhīd*, abū Rīdah edition, beginning.
70. The first question that has to be settled in *Tahāfut al-falāsifa* is whether the world is *qadīm* or *muḥdaṯ*.
71. Ibn Suwār, in his treatise 'Demonstrating that the Argument of John the Grammarian Proving the Contingency of the World is More Acceptable than the Argument of the Theologians' (translated by B. Lewin in *Mélanges, Nyberg*), distinguishes three meanings of *muḥdaṯ*: (a) something the existence of which, starting from an origin, is completed 'in a space of time and its revolution'; (b) something which, like sensation or intellectual perception, 'has no time'; (c) the effect which is contemporary with its cause, the latter having over the former 'a priority of nature and rank'.
72. cf. *M.I.*, Vol. I, p. 227 et seq.
73. Maimonides, *Guides Des Égarés*, Vol. I, p. 313–4.

74. *Manāhij-al-'adilla*, p. 135 et seq. This intention is very clear in the *Tamhīd* of Bāqillānī.
75. M. Guéroult, 'Physique et métaphysique de la force chez Descartes et chez Malebranche', *Etudes sur Descartes, Spinoza, Malebranche et Leibniz*, New York, Olms, 1970.
76. Attention should be drawn to an essential difference between Descartes' theory of discontinuous time and the theory of Muslim theologians: in Descartes, the discontinuous nature of time *requires* the infinite divisibility of matter; for the Muslim theologians, the discontinuous nature of time goes side by side with, and is required by, a 'material' atomism. cf., with regard to Descartes, M. Guéroult, *Descartes Selon l'Ordre des Raisons*, Vol. I, p. 281–2, Paris, 1968, and also J. Wahl, *L'Idée de l'Instant dans la Philosophie de Descartes*, Paris, Vrin, 1953.
77. The axiom 'the whole is greater than the part' is also valid for the infinite.
78. Aristotle's definition (cf. *Physics* IV, 219^b, 1–2).
79. Kindī, op cit., p. 95–7.
80. For Aristotle, the recognition of one kind of infinity in cosmology is 'indicative of the fact that magnitude as a field of operations (characterized by the additivity of all elements of magnitude whatever) is consciously and conceptually distinct from magnitude represented as a physical dimension' (J. T. Desanti, *Une Crise de Développement Exemplaire*) and it is thanks to this distinction that the operational field of *mathesis* can be treated as autonomous. In the case of Kīndī, are we to conclude from the stratification of meanings which is characteristic of the term *jirm* (which is, simultaneously, *body*, *substance* and *magnitude*) that no such distinction exists? This question would require separate study.
81. Aristotle regards movement also as a measure of time (cf. *Physics* IV, 12, 220^b–1432).
82. Kindī seems to be making a contraction of Aristotle's and the Stoics' definition of time (and even Plato's) according to a tradition reported by Plutarch in *De Placitis Philosophorum*, I, 27. cf. Wolfson, 'Crescas' Critique of Aristotle's *Physics*', in *Jewish and Arabic Philosophy*, p. 636–9, Cambridge, 1929.
83. Kindī, op. cit., p. 97–8.
84. J. Vuillemin, *Cinq Etudes sur Aristote*, p. 130.
85. Kindī, op. cit., p. 98, 99.
86. ibid., p. 99.
87. *Tahāfut at-Tahāfut*, p. 153–4.
88. Kindī, op. cit., p. 99–100.
89. cf. Ibn Rušd, *Tahāfut at-tahāfut*, p. 152. On Ibn Rušd's solution to the problem of the eternity of the world see: Mājid Fakhry, 'The Antinomy of the Eternity of the World', in 'Averroes, Maimonides and Aquinas', *Le Muséon*, 1953.
90. *Manāhij*, p. 136.
91. *M.I.*, Vol. II, p. 116 (question 148).
92. This is merely a paraphrase of *Risāla* by Qušayrī, p. 52–3, 1966 edition.
93. ibid. p. 69.
94. For a description of this experience see the study by R. Arnaldez, 'Dynamique et Polarité des Etats Mystiques', *L'Ambivalence dans la Culture Arabe*, p. 143–52, especially p. 148, a collection edited by J. P. Charnay, Paris, Anthropos, 1967.
95. Quoted by Qušayrī,
96. Junayd, cf. Sarrāj, *al-Luma'*, p. 212. Quoted in Arberry, *Sufism*, p. 59, London, Allen & Unwin, 1950.
97. M. Foucault, 'L'Ordre du Discours', inaugural lecture at the Collège de France, 2 December 1970.
98. M. A. Sinaceur, 'Connaissance des Arabes', *Critique*, Vol. XXVIII, No. 298, March 1972, Paris.
99. P. Burgelin, *L'Homme et le Temps*, p. 109, Paris, Aubier, 1945.

TIME, TEMPORALITY AND FREEDOM

Seizo Ohe

Dōgen's reflections on time

Time is existence and all existence is time. ... Because time's transit leaves traces, man does not doubt it. Though he does not doubt, he does not understand. Because the ordinary man does vaguely doubt everything he does not understand, his future doubts may not agree with his present doubts. Even doubt is nothing but a part of time. There is no world without this doubting self, for this self is the world itself. We must regard everything in this world as time. ... Thus we see that the self is time itself. ... And each grass and each appearance are time.*

This citation is taken rather freely out of the opening part of an article on '*Uji*' (literally translated: 'being time') in the famous *Shōbōgenzō* by Dōgen (A.D. 1200–53) who transmitted Sōtō Zen to Japan around A.D. 1230.

Obviously time is considered by Dōgen not as a form of consciousness but as consciousness itself and our own conscious self is further identified with everything we are conscious of. The effect of this all-identifying time-experience is furthermore accentuated in the all-embracing experience of each moment, as is evident in the following sentence from the same passage: 'In each moment there are all existences and all worlds.'

But Dōgen's philosophy of time is ultimately rooted in his own

* Throughout this article the author freely uses—with permission of the publisher—Professor Reiho Masunaga's English translation (with some modifications) of Dōgen's Japanese text: *The Sōtō Approach to Zen*, Tokyo, Layman Buddhist Society Press (Zaike Bukkyo Kyokai), 1958. So far this is the only source of the *Shōbōgenzō* available in a Western language. It is now unfortunately out of print. The *Shōbōgenzō*[1] is a collection of ninety-five articles written by Dōgen during the latter half of his life, after his return from China in 1228.

immediate time-experience. So he dwells on it more concretely and more personally in the following passage:

> When climbing a mountain or crossing a river I was present, and if I am, time is. As I am here now, time cannot be separated from me. If time does not have the form of coming and going, the moment of climbing a mountain is the eternal now. If time takes the form of coming and going, I have the eternal now.

In another article of the *Shōbōgenzō* entitled '*Daigo*' (literally translated: great enlightenment), Dōgen expands this rather specific exposition into a universal observation on man's time-experience in general: 'The so-called present is everyman's now. When now we think past and future, myriad years are present. They are the now. The original nature of man is the present.'

The moment 'here now' is the eternal present that embraces all existences and all worlds in its experiential plenitude and timeless continuity. Dōgen then contrasts this enlightened time-experience with the common concept of time:

> Most people think that time is only transitory. They do not understand that coming time is also time and each being time dwells in its own situation. Even if recognizing that time always dwells in its own situation, who has the freedom to express it? Even if you could express this attainment over a long period, you would still be groping for its natural truth. If you think of time in the common way, even wisdom and enlightenment become mere appearances in time coming and going.

Toward the close of the article on '*Uji*', Dōgen repeats, using all possible modes of expression and association, the doctrine of the self-identity of time and things with which he began his reflection on time in this article, so that we may become in the end entirely convinced of the fundamental truth that not a single thing can arise apart from time, and that enlightenment can be realized for well-trained minds in the full experience of this identity with all things at every moment everywhere. Nirvana is here now.

This is a crude outline of Dōgen's philosophy of time as set forth in the article entitled '*Uji*' of the *Shōbōgenzō* which he wrote in 1240 when he was forty-one years of age.

Dōgen's philosophy of time and its core wisdom

Dōgen's philosophy of time with its thoroughness and depth of thinking stands out conspicuously even in the whole history of Buddhist thought—Indian and Chinese as well as Japanese—where the problem of time has always played a part in the problem of the temporality of human existence and its overcoming by enlightenment.

On the other hand, as we have seen in the previous section, Dōgen knew very well the epistemological primacy of time, just like Kant, because he knew that our cognitive consciousness does rise and fall with our time-experience. But he went further intuitively, like Bergson, to look into the metaphysical realm of time-continuity where man's spiritual energy erupts, staggers and stands still or violently runs from moment to moment. Dōgen did not stop here. He went deeper (as did Heidegger or Husserl), into the very basis of our experienced time, where the essence of human existence may reveal itself at any moment. Even then he did not stop. Dōgen proposes to all truth-seekers to sit cross-legged quietly and try to think the unthinkable, i.e. to keep the mind as empty as possible of all the illusions and delusions of everyday life, by self-identification of body-mind and by the harmonious control of breathing. Only by this practice of *Zazen*, he thinks, can man attain true enlightenment in a timeless void of the 'eternal now', overcoming the temporality of human existence.

I myself have for many years been unable to understand why Dōgen emphasized the *Zazen* practice so much, recommending it to all truth-seekers with as much confidence as enlightenment itself. Just a few months ago, however, I came to understand it when I was casually experimenting with *Zazen*, sitting in a half cross-legged manner—a *Zazen* sitting position much easier and stabler for laymen than the full cross-legged one—harmoniously controlling respiration. The respiratory function being one of the most vital physiological functions of the human body, one that can partly be put under the voluntary control of the human mind, I suddenly began to feel somehow as if my body and mind were fused together into one, which is most probably the original state of the human organism itself, as it was created in nature thousands of years ago, and as may still be the case with very small children today.

How did it come about that this original, natural harmony between body and mind in the human organism is so miserably crippled in most adult human beings today? Klages was right in blaming *Geist* as the *Widersacher der Seele*, but wrong in not adequately seeing the original harmony of all the organic functions of man. Through all kinds of educational inculcations appropriate only for their respective social environments, men of civilized society today are becoming like low-speed computers with definite set programmes, losing more and more of their inborn natural flexibility in thinking and action, thus strengthening the sense of temporality of their own existence. Yet most of them are internally suffering from it without knowing the cure. Many people are interested in Zen throughout the world. Dōgen's Zen in particular attracts people by its strong emphasis upon the *Zazen* practice which by itself attempts to help them restore the integrity of the human organism, to re-establish the lost harmony of body with mind, of man with nature, and to overcome the temporality of their own existence.

Dōgen's philosophy of time has thus led us to the somewhat mysterious problem of *Zazen*. Let us try to look into the core of this Zen wisdom and explicate it in as plain words as possible.

To do what is right, we all know, it is not enough to know what is right. We do only what we would like to do. Therefore, in order to do the right thing, we have first not only to know it intellectually but to want it actually. As a rule, we are encouraged in action by joy or pleasure and discouraged by pain or sorrow. This means the mechanism which ultimately controls human action is not knowledge or reason, but desire or will, which are commonly involved in different degrees with bodily feelings. We cannot do anything without emotional support of some kind or other—just as we cannot do or even know anything without time-experience (which is always in the present). There is no present, on the other hand, apart from human experience, be it by knowledge or action. It must be here, now that we can live the true life with our whole body-mind in genuine joy. *Zazen* is just the way of re-establishing the original body-mind mechanism of human behaviour in its natural integrity. It is the way of living the truth and seeing the true self in a unique time-experience of the eternal present. The sorrowful sense of temporality which always accompanies human existence will thus be overcome. We will be free not only from the worries of every day but from illusions and delusions of the separate self, of man and nature against each other, of the divided body and mind, etc., in a naturally welling-up feeling of profound pity for all beings on earth.

This is precisely what Dōgen sums up in the following dictum of *Genjōkōan*[2] of the *Shōbōgenzō*:

> To study the truth is to study the self.
> To study the self is to forget the self.
> To forget the self is to be enlightened by all things.
> To be enlightened by all things is to be free from attachment to the body and mind of one's self and of others.

It may be in the same vein of 'Japanese spirituality',[3] to borrow the late Dr D. Suzuki's words, that even in the continuous succession of *Nembutsu Shōmyō*, *Namu-Amida-Butsu* (adoration for Amida Buddha) of the Pure Land Buddhism, the instant of 'one thought' is stressed with each calling of the Name, although here the attitude is completely passive, entrusting the self wholeheartedly to the Other-Power (*Tariki*) of Amida Buddha, whereas in *Zazen* the attitude is fundamentally active, relying solely upon the Self-Power (*Jiriki*) of one's own inborn Buddha-nature. The same character of an instantaneous all-energetic act of faith is accentuated in the *Ōchō* that was required by Shinran (1173-1262), the founder of the Shin (i.e. True Pure Land sect), as being absolutely one's rebirth in the Pure Land.

It is very characteristic for all these instantaneous acts of religious conviction in Japanese Buddhism to have an immediacy which is at bottom emotional or even sensory. An overwhelming feeling of plenitude throughout one's whole body-mind is common to all of them. There is no room for the sorrowful sense of temporality that always accompanies human existence—at least for that moment. For the contemporary Japanese intellectuals, however, *Zazen* is obviously more attractive and significant than *Nembutsu* is. Therefore, it is the tradition of the former rather than the latter that thinking minds of modern Japan have been following in their incessant efforts to establish self-identity amidst foreign cultures coming from all over the world, and in particular the modern Western civilization of science and technology which they cannot dispense with at all.

Naturally, it is also along this line of the Zen tradition that we are going to proceed in our subsequent observations on time, temporality and timeless void.

Time, temporality, and timeless void

Since that which we call time is nothing more than a form which orders the content of our experience, for the idea of time to arise, the content of consciousness must first be able to be fused, to be united, and to become one. If this were not so, we would not be able to think temporally, linking and ordering that which takes place before and after. Thus, the unifying function of consciousness does not receive the control of time, but on the contrary time is established by this unifying function. We must say that at the foundation of consciousness there is a certain transcendent, unchanging thing beyond time.[4]

This is a passage from *A Study of Good*, by Kitaro Nishida (1870–1945), one of the most representative philosophers of modern Japan. There is no doubt that the position Nishida took at this initial stage of his 'philosophical pilgrimage'[5] was not much different from those of modern European and American philosophers who wanted to defend man's spiritual reality with its proper degree of freedom against the ever-advancing causal-deterministic explanation of all things, internal as well as external, by science, which tries to spatialize time in one way or another. This is a position in regard to human consciousness or immediate experience, with time as its intrinsic form.

This initial position of 'pure experience' that Nishida developed in his first work, however, was already rooted in his personal *Zazen* experience of many years, as we see clearly in the last sentence of the above citation. It seems to have left in him a lasting impression and urged him to give it a philosophical expression such that he could intellectually realize the spiritual calm and strength of a Zen master in the midst of this modern world of science and technology. Finally after some twenty years of philosophic reflection Nishida came to attain with his third and most remarkable major work, *From the Acting to the Seeing* (1927), the unique idea of a 'place of nothingness' where the creative activity of the true self comes from and returns to—a timeless void, as it were, wherein everything can be placed. But this is a philosophy too placid for us to live up to in the historical reality of our time. It may well be said to have something of the spiritual calm of a Zen master but nothing of his actual strength, as we find in such contemporary Zen personalities as the late Dr Daisetz Suzuki and Dr Shinichi Hisamatzu of the FAS Zen Institute in Kyoto.[6]

After all, philosophy is a work of intellectual contemplation, while Zen is practical training for establishing a body-mind mechanism flexible enough for coping with anything. Zen must help us, on the one hand, to accept calmly the temporality of our existence as such with placid contemplation in *Zazen*, but, on the other hand, to act creatively in any work and at any moment of our daily life.

This last point is just the key to another unexpected development of the Zen tradition in the contemporary Japan of advanced technological industry. If Zen, in particular *Zazen*, is really capable of establishing a new body-mind mechanism in man beyond time and re-establishing the harmony between mind and body as given to man thousands of years ago after the accumulating biological evolution for millions of years of the earth's history, then it must also be possible for every human being to revive by *Zazen* his original creativity which is now lost or lying dormant in him for most of the time under the pressure of heavy educational and professional routine work. Therefore, it is no wonder that the creator of the so-called NM Method[7] in creative engineering has recently come to co-operate with a Zen practitioner and is now developing a unique technique of creativity training for engineers. The technical details aside, what interests us most is that the time-transcending orientation of the creative moment, 'here now', in *Zazen* seems to be playing an important part in this technique, so as to reshuffle the highly mechanized intelligence of a civilized man into its original state of natural flexibility, and so, with appropriate emotional support, inducing human intelligence to creative thinking.

Creativity is certainly one of the most distinctive characteristics of human intelligence. But the time-bound routine work of our daily life makes little use of it, only some exceptional men of genius make the best use of it, at some time-transcending creative moment. Creative ideas in art and science often appear suddenly in the human mind, just as Zen enlightenment does, in an instant, though perhaps after long years of contemplation. 'If you want to see, see right at once. When you begin to think, you miss the point.' So a Zen master used to teach his disciple.[8] This time-transcending instantaneousness is characteristically common to Zen enlightenment and to creativity in art and science. It emancipates human beings from their attachment to body and mind, to pleasure and pain, to life and death, or to fame and riches, at least momentarily. We definitely need such moments of genuine joy in our life from time to time. Creativity in art and science is a sort of enlightenment in our intellectual life.

Time-consciousness and human freedom

Philosophy must cope with the reality of the present age, the philosopher with the possibilities of the near future. Since nowadays science is showing us 'the exciting possibility of tracing the pathways from interstellar matter through the first informational macro-molecules and primitive cells to the human brain',[9] we have to locate the genesis of time-consciousness somewhere, though certainly on a very advanced level, along the long evolutionary line of matter, life and mind.

It is amazing that both science and Zen have in common the intellectual courage to face reality at all costs, though science sees reality objectively, Zen subjectively.[10] For instance, both science and Zen in principle acknowledge the oneness of matter, life and mind, especially the inseparability of body and mind in human beings. As for time, however, while science usually sticks to the physical time measured by some movement of matter (although some life processes might need another way of time-measurement), Zen knows only experienced, subjective time. But as far as time-consciousness is concerned, science as part of the creative activity of human intelligence cannot of course exist apart from it. As Dōgen identified it with experience itself and even with experienced objects themselves, so time-consciousness always accompanies any living human being, including the working scientist, at times tormenting him, at times boring him. At its most creative moment, however, some splendid new idea catches the scientist at work and some joyful feeling overtakes his whole body-mind, annihilating the time-consciousness as such. It is a moment of time-transcending creativity. Such a moment of free creation occurs in art and science, just as a moment of enlightenment occurs in Zen, particularly in *Zazen*. But such a practical possibility of instantaneous freedom must have a deeper theoretical foundation.

In fact, science is now teaching us from its viewpoint of objective cognition the fundamental continuity of matter, life and mind, suggesting with some experimental evidence that there could have existed a certain protenoid microsystem which was 'a common ancestor of all life on earth',[11] and that mind as well as life should calmly be regarded as nothing more than a highly elaborate function of an extremely complicated organization of matter which is formed by the accumulating process of chemical and biological evolution at a certain step from a lower level to the next higher.

But, on the highest level of this whole evolutionary process, enlightened science of today does not hesitate to recognize the human organism as fully capable of the so-called spiritual activity with its proper degree of freedom which is, however, commonly considered by science as much more limited than Zen traditionally will have it.

Indeed human intelligence as the bearer of man's creative activity in science must have some degree of freedom, but human freedom is not in any event as great as Zen always seems to believe. As a matter of fact, man's spiritual activity, be it scientific-intellectual or moral-emotional, is so narrowly limited, first of all by heredity, then by natural environmental conditions on the one hand and by socio-cultural conditions on the other, that human freedom appears, after all, to be nothing but an illusion. Nevertheless, man is in fact free within these narrow limits of all the determining conditions. Otherwise there could not have been such great works of art and science as we see all along the history of mankind. They bear unmistakable evidences of human freedom.

To repeat, man's freedom is not great, but without this bit of freedom he has, there is no life worthy of man. Now we all know that science and Zen are trying to enlarge this freedom of man, each in its own way: science by objective knowledge, Zen by personal exertion. To heat the room in cold winter and cool it in hot summer, that is the way of science. To train body and mind so as to stand any heat and cold, that is the way of Zen. We need both ways. To remove through science and technology external causes which make us suffer is no less important for our existence than to attain the internal strength which gives us immunity to any suffering. Science and Zen should now be integrated into the whole of human wisdom. The future destiny of mankind will largely depend on its success.[12]

On the whole, science today stands surprisingly close to Zen; both respect experience and see reality as it is. To use Dōgen's phraseology, both are 'studying the truth', but science is more 'forgetting the self', while Zen is more 'studying the self'. They both are being 'enlightened by all things' and so 'free from attachment to the body and mind'.

It may be of some interest to add that several years ago a Japanese authoress wrote an imaginative drama[13] whose hero is a man not only free from attachment to the body but entirely emancipated from the body itself, i.e. a purely spiritual, time-transcending being named Emiton ('no time' spelt backwards).

NOTES

1. The *Shōbōgenzō* (The eyes and treasury of the true law) is the life-work of Dōgen, the founder of Japanese Sōtō Zen, who is now regarded as the most profound

philosopher among all the Zen masters of Japan and China. It is attracting more and more attention from the general public in Japan. Since the original text is not easy to read, three editions of it in modern colloquial Japanese are being published at the same time: Masanobu Takahashi's in two volumes (completed) (Tokyo, Risosha, 1972); Sōichi Nakamura's, with original text and commentaries, in five volumes (completed) (Tokyo, Seishin-Shobō, 1973); Fumio Masutani's, with original text and commentaries, in eight volumes (not yet completed) (Tokyo, Kadokawa-Shoten, 1973). Besides Professor R. Masunaga's English booklet which I have mentioned at the very beginning of this paper with special acknowledgement, the following two may also be of use to the Western-language readers for general orientation in Zen, though the translated texts are mostly of Rinsai Zen: M. Shibata and G. Renondeau, *Le Sermon de Tetsugen sur le Zen*, Tokyo, Rishosha, 1960; S. Ohasama and A. Faust, *Zen. Der lebendige Buddhismus in Japan*, Gotha-Stuttgart, Friedrich Andreas Perthes, 1925.

2. English translation by Reiho Masunaga, op. cit., p. 126–32 (see note on page 81).
3. Daisetz Suzuki (English trans. by N. Waddell), *Japanese Spirituality*, p. 136, Tokyo, Unesco, 1972.
4. Kitaro Nishida (English trans. by V. H. Viglielmo), *A Study of Good*, p. 63–4, Tokyo, Unesco, 1960.
5. Seizo Ohe, 'Modernization and Japanese Philosophy', *The Transactions of the Asiatic Society of Japan* (Tokyo), Vol. 3, No. 9, 1966, p. 77. See also, Seizo Ohe, 'Japon', in: M. F. Sciacca (ed.), *Les Grands Courants de la Pensée Mondiale Contemporaine—Panoramas Nationaux*, Vol. II, p. 943–7, Milan, Marzorati, 1959.
6. FAS stands for the three goals of the institute: to awake to *F*ormless self; to stand on the standpoint of *A*ll mankind; to create *S*uper-historical history.
7. NM stands for Nakayama Masakazu, the full name of the originator of the method (in Japanese order). Professor M. Nakayama is currently the president of the Creative Engineering Institute Inc., in Tokyo. The Zen master who co-operates with him is Tazato Yakumu of Shaka-dō, Kamakura.
8. Daisetz Suzuki, *Zen and Japanese Culture*, p. 13, New York, Pantheon Books, 1958.
9. Sidney W. Fox, 'Evolution from Amino Acids: Lunar Occurrence of Their Precursors', *Annals of the New York Academy of Sciences*, Vol. 194, 1972, p. 84.
10. Daisuke Ueda, *Zen and Science*, p. 73–83, Tokyo, Risosha, 1963.
11. Fox, op. cit., p. 83.
12. Seizo Ohe, 'La Philosophie Ouverte et la Sagesse Orientale', *Revue Internationale de Philosophie* (Brussels), No. 93–4, fasc. 3–4, 1970, p. 600.
13. Yachiyo Kato, [*Emiton*], Tokyo, Risosha, 1966.

Sociological interpretations of time
Pathology of time

SOCIOLOGICAL INTERPRETATIONS OF TIME AND PATHOLOGY OF TIME IN DEVELOPING COUNTRIES

Honorat Aguessy

In fact, the reason why it is possible, by bringing together several individual consciousnesses, to put the various thoughts or events which they contain back into a common time or times, is that internal duration can be broken down into several streams, originating in the groups themselves. Individual consciousness is only the place through which these streams pass, the point at which the collective times meet.—Maurice Halbwachs, *La Memoire Collective.*[1]

The problems raised by time reflect the great diversity not only of philosophies, which differ from one thinker to another, but also of the ways in which collective time is viewed, and change from one culture to another.

To appreciate the contribution of the sociological interpretation of time properly, a glance must be taken at the difficulties left unelucidated by pure conceptualization in regard to time.

First, it is difficult to conceive the world as presence of things in any of the variations on the 'cosmological' view (Aristotelian analysis, Galilean epistemology, or even Einsteinian relativity). Millennium concepts pose a problem. Some difficulties can be eliminated by taking a subjectivist view, i.e. by considering not the cosmos but the subject, but others arise instead. Ranking the inner experience of time above any other implies experiencing it at a more basic and deeper level, and defining the sources of this inner experience, i.e. the thinking processes which accord pre-eminence to one of a series which includes, for example, mechanical time. These processes are themselves pointers to a more deeply rooted experience of time. What then is the nature of this experience through which an ill-defined secondary level of consciousness is mediated? Leaving aside philosophies of concept (Kant) and existence (Sartre), and depending only on phenomenology (that 'sensitive method for studying temporality'), I notice that it is difficult for me, as an ego, to understand that I belong to immanent time and am the absolute source of meaning. In other words, how can I attribute time to myself, and attribute this time to myself within time?

Referring to Husserl's analysis in *Cartesianische Meditationen* and *Zur Phänomenologie des Inneren Zeitbewusstseins*, Jean T. Desanti wrote: 'How can fundamental time be simultaneously the terrain in which the experience originates (its first soil), and the field in which its meaning emerges and which in a totality teleologically organized for an Ego, offers the history of the experience?'[2] 'The philosopher as he proceeds experiences the fact that he has neutralized neither time as a link with the world sharing the same origin as the Ego, nor others, as a presence making effective for themselves the meaning of the Ego's own experiences....'[3] He concludes: 'Self-awareness is no longer to be sought in the proud solitude of the absolute thinker: each attains it through the mediation of others, through the hard work which, discovering the meaning of present history, endeavours to materialize it as a task and express it as a truth.'[4]

What, first, do men experience together before the individual accumulates his own experience of time—since it is within a given culture that each first experiences time? We must obviously seek an answer from different cultures.

Let us first consider how different societies appear from the point of view of their characteristic conception of time. It is said that no one can escape from his time, i.e. each society is apprehended in terms of the time which seems to mark all its institutions and activities. Taking the classification of societies commonly used by economists, a characteristic time could be ascribed to each, e.g. pre-industrial and non-industrial (rural) societies are preoccupied with the present, while industrial and post-industrial countries are stimulated by the future.

For pre-industrial and non-industrial (rural) societies to be characterized by the weight of the past, their activities must be purely repetitive; they should rely only on ancestral tradition, never innovate, never look beyond what nature provides, never modify it in any way. The same sequences would be repeated, there would be political unanimity and social consensus. Intolerance of the new (misoneism) is, according to Lévy-Bruhl, an essential trait of such a society, which rejects any outside social element likely to set off a process of disintegration. 'They resemble organisms which can live for a very long time, provided their external surroundings vary little, but which, if new elements break in, rapidly degenerate and die.'[5] Misoneism thus demands a meticulous, exclusive respect for the ancestral attitudes to time: 'The supreme rule is thus to do what the ancestors did, and nothing else',[6] so that 'even an established innovation remains for a long time precarious'.[7] The function of traditionalism, as 'a social usage which regulates behaviour' is to 'encourage conformity and maintain as best it can the "repetition" of social and cultural forms'.[8] The weight of the past thus seems more important than any other dimension of time. The primary concern is with the transmission of codes and traditions.

Let us now consider the characteristics of consumer societies which would justify their preoccupation with the present. The act of consumption becomes more important than the activity of production; the consumer is considered 'as an outsider subjected to a decision-making system operated in the name of the community'—as if, acting on a certain conception of progress, he no longer needs to refer to tradition or history: the economy is the essential point of reference. With regard to time, we learn from the many studies in recent years on the consumer society that the consumer, browbeaten by the mass media and high-pitched advertising, keeps within the temporal nucleus of an ever-renewed present imposed on him by society. Having no links either with the values of the past (a way of life or a culture) or with a historical perspective, this society is concerned solely with the present. As Alain Touraine writes:

> The consumer has usually only very little control over the producer. The only sanction he wields is the refusal to consume, and even this is weakened as the consumer's attitude becomes more passive; in regard to the cinema, for example, when the content of the film is less important than just regularly getting away from the everyday in a darkened hall.[9]

Rejection of the past as a burden and encumbrance, and the absence of any general perspective on which to base the future should enhance the significance of the present. This does not in fact seem to have happened in industrial and post-industrial countries. Touraine in *La Société Post-industrielle* uses three adjectives: 'post-industrial', 'technocratic' and 'programmed'. Here it might be preferable to describe post-industrial societies as rapidly changing societies, and societies in which acquisition is more important than transmission. The rapidity of change, which leads to new creation, really stimulates the future. Touraine notes 'the importance of reasoned forecasting based on statistical laws and generalizations concerning man and the group, or in short, prospective studies and programming'. Growth is a valuable indicator of change. As Touraine says:

> Growth is the outcome of a number of social factors taken together rather than of the mere accumulation of capital. The most novel development is that it depends much more directly than before on knowledge, hence on the society's ability to create creativity. Regardless of the aspect selected—the role of scientific and technological research, vocational training, the ability to programme change and control the relationship between component parts, to manage organizations and hence systems of social relations, or to encourage attitudes that favour the mobilization and continuing transformation of all the factors of production—all aspects of social life, education, consumption and communication are more and more closely integrated with what might formerly have been called the forces of production.[10]

Thus the options taken on the basis of these various factors of change 'transform the indefinite future along precisely determined lines'.

Controlled changes rapidly overtake current techniques, which soon become obsolescent unless change is slowed down by the defence of vested interests. Ways of life change equally rapidly. Time is accelerated and the future affects or even determines the present. Georges Balandier recalls that such 'enterprising societies' are subject to increasingly numerous and rapid changes, and to breaks so sharp that it seems as if the present can be kept under control only by keeping constant watch on the future. The 'prospective' approach is the technical (or technocratic) reply, and 'futurology', which originated in neo-capitalist countries, is spreading to the socialist countries. To explain this proliferation of 'futurologists', regardless of ideology, Balandier added: 'This is the first time in its history that mankind has seen so far-reaching a series of mutations in the space of one generation.'

These accordingly are some of the characteristics of the three types of society to which a particular dimension of time is exclusively attributed: past, present or future. From this point of view it is quite possible to look on any attitude which is out of line with these time dimensions as pathological, e.g. addictions to the past underlying modernism; the 'protest' generations, obsessed with the instant now; or the archaism and immobilism of pre-industrial societies. A lot can be learned by checking the uniform attitude to time ascribed to each of the three types of society with its supposed 'pathology'.

Is this so-called 'pathology' not in fact simply one constituent aspect of the dynamics and complexity of each society's attitude to time? In other words, does 'pathology' not so much describe what must be expelled in order to achieve health as indicate what constitutes health?

Durkheim described as pathological any institution which no longer served a need, and yet survived after its reason for existing was gone. Communication is one of the factors involved in the sociological approach; can the attitude to time of a given society be studied if those elements which are considered pathological are rejected?

The continuity hidden from the careless (or those obsessed with some single aspect of society) by the increasing speed of change in certain sectors is certainly part of the dynamics of any society. In this sense an addiction to the past underlying modernism does not transfigure modernism, but reveals its true face. It is paradoxically because of the breathless rate of change in certain sectors that the past holds undivided sway in others. Writing of the survival or resurgence of addiction to the past beneath modernism, Maurice Halbwachs says:

> There is hardly any society we have lived in for any length of time that does not survive, or at least leave some trace, in more recent groups in which we have mingled: the survival of such traces suffices to explain the persistence and continuity of the time peculiar to that earlier society, and the fact that we can at any time re-enter it in thought.[11]

This former student of Bergson sees collective times as the same, regardless of the speed or slowness of change:

What distinguishes these collective times is not that some elapse more quickly than others. It cannot even be said that they elapse at all, since each collective consciousness can recollect, and memory does indeed seem to be contingent upon the survival of time. Events follow each other in time, but time itself is a fixed framework. All one can say is that different times have different extensions, and enable memory to go more or less far back into what is conventionally called the past.[12]

In other words, it is not possible to cast out of time those of its aspects which might be described as pathological. By showing the limits of the 'normal', the 'pathological' reveals the rich complexity of time.

What industrial and post-industrial societies call pathological (given the speed of change in gadgetry, processes, machinery) are matters of individual and collective experience, regional sentiment, and so on. The clash is no longer in attitudes to time proper to different cultures or civilizations, but primarily in those within a given culture or civilization. Societies which to themselves seem deeply divided regarding their headlong changes appear to other observers or analysts to be imbued with normal traditions which no mere public pronouncements can uproot or obliterate. Balandier writes:

The intensity of change may mask the evidence of continuity. However, certain current expressions show up these permanent features by identifying them as obstacles to modernizing forces, e.g. expressions which complain of the sociological heaviness and inertia of certain kinds of behaviour, or affirm the need for structural reforms. ... For it is important to differentiate clearly the incidence of tradition at any particular stage reached in growth process. In western Europe, during the period of laissez-faire accumulation which led to the greatest tensions between the social classes, tradition was invoked as a means of reinforcing social distances and barriers—as has been noted by writers from de Tocqueville to S. M. Lipset.[13]

This explains the continuing co-existence in industrial and post-industrial countries of modernism (which they regard as their own) and traditionalism (the time designations which they assign disdainfully and exclusively to pre-industrial and non-industrial societies). The prestige and sense of community evoked by constant reference to a glorious past redolent of ordeals surmounted, and to today's high culture, reflect the importance they attach to striking a balance between past, present and future, and indicate that they have no intention of concentrating on power and the future to the exclusion of all other values. The addiction to the past that underlies their modernism denotes a complex social dynamism.

Are pre-industrial societies really archaic? If so, an analysis should reveal specific disparities in the attitude to time of their constituent sectors. Unfortunately, as usually described and criticized, they tend to be the imaginary opposites of real industrial societies, an ideal type (cf. Dilthey) which acts, in an analysis, as a pole of repulsion. 'The concept of a traditional society represents a reversal (the opposite of the industrial society) rather than a sociological type.'[14]

Exclusive emphasis on the past in discussing 'traditional' societies would already seem unjustified to the sociologist or historian. If, in addition, we further emphasize addiction to the past by saying that these societies are archaic and immobile, we merely strengthen the negative features of a sociological monstrosity. To the sociologist or any informed person, the attitudes to time of these societies are quite as diversified as those of other types of society, and vary at economic, political, religious and other standpoints.

Let us revert to Lucien Lévy-Bruhl's comment, quoted above, on the misoneism of traditional societies. After stating categorically the pathological archaism that 'the supreme rule is thus to do what one's ancestors did, and nothing else', he later says:

In fact, there is hardly any society so lowly that it has not yielded some invention, some industrial or artistic process, some admirable fabrication: dug-out canoes, pottery, basket-work, textiles, ornaments, and so on. The same men who, lacking almost everything, seem to be on the lowest rung of the ladder, will achieve a surprising delicacy and precision in the production of a particular object.[15]

The question then is whether the ability to produce such objects is also handed down from generation to generation: acquisition or transmission? If transmission, how far must we go back to trace the original appearance of the object, tool, instrument or institution supposedly inherited from ancestors? This question the proponents of the misoneist argument never broach, much less answer. Instead, inspiration led certain 'scholars' to propose an abstract framework in which immobilism is ascribed to traditional societies. But any sociologist dealing with 'traditional societies' knows that change, the creation of tools to meet new needs, the ability to foresee the immediate or future consequences of present actions, are common to all cultures; none lives in a one-dimensional time facing only towards the past.

There are thus certain attitudes to time specific to each of the sectors of a society; and a clash, or at least a disparity, between these different attitudes. On the sum of these different attitudes depends the driving force of a society; the apparently 'pathological' features are but another of the normal conditions for social health, which are overlooked by an incomplete, biased approach. More specifically, the driving force of a society should not be confused with the attitude to time of any single sector of that society, whether economic, political, religious or other: it depends on the interplay of all of them. And it is determined in each society by the unique way in which the different processes and elements interlock.

This being so, what is meant by the 'pathology' of universal time? How is it that many thinkers are tempted to situate the collective times of different cultures in relation to a homogeneous time and a history with a specific orientation?

One day it may perhaps be possible to show that all societies tend to see their own attitudes to time as the only valid yardstick. As it is, many thinkers from countries which have dominated others are tempted by the idea of a universal time which establishes the perspective for all the others. They identify the present of others with their own past; they jumble together events that were of far-reaching significance to those who experienced them; they treat as rigidly immobile something which has its own subtle dynamics. What to some people is life and spirit is reduced to matter, matter regarded as the *mens momentanea* of Leibniz.

Let us try to outline briefly some of the theses based on the idea of universal time.

Hegel represents universal time as an ever-flowing stream, bearing away corpses which are backward societies and depositing them at its bends. Hence it is not surprising (at least in Hegel) that the author of *Die Vernunft in der Geschichte* regards the attitudes to time of non-European societies as a demonstration of the cunning reason uses to reach the state of Europe after Napoleon's conquests. Telescoping or rejecting the attitudes of African societies, Hegel writes, on the subject of Negroes:

Their condition allows of no development or education. They have always been such as we see them today ... Whereupon we leave Africa behind, and shall not mention it again, since it does not form part of the historical world; it shows neither movement nor development, and such events as have taken place there, that is to say in the north, relate to the world of Asia and Europe ... In short, what Africa means to us is an a-historic, undeveloped world, entirely a prisoner of the spirit of nature, whose place is still on the threshold of universal history.[16]

Having thus set Africa aside, Hegel says of Asia that he finds himself at last 'in the true theatre of history'. However, it is not long before he rejects Asia as well: 'China, India and Babylon became great civilized countries. But they remained inward-looking, and did not adopt the seafaring principle, or at least not after their own particular principle had attained its full development.'[17] Asia, he adds, 'was the birthplace of all religious and political principles, but it was only in Europe that they developed'.[18] Lastly, his reasoning leads him to the following conclusion: 'In this way the life of the European States acquired the principle of the freedom of the individual', a final stage towards which all mankind's hesitant and stumbling steps were directed.

But Hegel is not alone in arguing thus. A very famous sociologist,

Max Weber, divides societies into two groups: societies whose attitude to time reflects their own dynamic rate of change; and societies whose attitude reflects the power of exogenous factors acting upon them. The first can programme and build the future, and act accordingly. The second are traditional societies, lacking internal dynamism, depending on the programmed societies to provide the impetus for change.

Rostow's unilinear pattern[19] is so well known that we need not dwell on the conception of time it reflects. Briefly, the economy (the only factor taken into account in his theory of the five stages through which all societies inevitably pass) imposes an attitude to time which impoverishes the dynamism of cultures.

These three theses suggest that all human societies of necessity follow exactly the same itinerary and develop the same procedures to reach the same destination. Are they valid today?

Dealing with the concept of universal time (covering all events that have ever occurred anywhere, all continents, all countries, all groups in each country and through them every individual), Maurice Halbwachs takes the question raised above from a different angle: 'The point of view of what group should be taken by an author who wishes to write the history of the world and escape from his own limitations?' He replies: 'The single time thus reconstituted includes vaster spaces, but it still covers only a limited part of the people concerned—the mass of the population which does not come within these limited circles but lives in the same places, has also had its history.'[20]

The study of different times within the same society is of such importance that any concept of time which does not take account of their multiplicity and complexity runs into insuperable difficulties.

Karl Marx was not unaware of these difficulties. He was invited by Vera Zasulich in February 1881 to clarify a crucial point in *Das Kapital*: whether he thought it absolutely essential that agricultural Russia must go through all the stages of capitalist industrial exploitation before it could hope for a revolution. In his reply of 8 March 1881, Marx wrote that the thesis applied primarily to Western Europe, and that conditions in Eastern Europe might be different. He believed that two solutions were a priori possible. After describing the development of the agricultural commune, which he called the first social grouping of free men not linked by ties of blood, he showed that such communes represented a transition between primary and secondary formations. He then posed the decisive question: 'Does this mean that, regardless of the circumstances, the agricultural commune will always develop the same way?' His reply leaves no doubt: 'Its constitution allows of the following alternatives: either the element of private ownership it implies will prevail over the collective element, or vice versa. It all depends on historical conditions and the background ...

Either is possible a priori, but each obviously implies an entirely different historical background.'[21] Marx thus demonstrates that processes are not mechanically linked within a society. This position sets him apart from the proponents of a single, universal time. In a certain sense it sets him apart from Engels, at least the Engels of 1852, for whom the Russian peasant commune was hopeless as a basis for socialism, since he believed there was only one possible route which all societies must follow.

The image of a unilinear current of one universal time against which the attitudes to time of different societies fall into perspective has a certain intellectual fascination. Thinkers aspire to the universal, and want to find a strict, coherent pattern in human affairs. But collective attitudes to time are not amenable to an analytical logic which will reveal the truth by simple deduction. They are a product of human communication which is pragmatic. The point at issue is not reason, surreptitiously determining who shall be master and who servant, but relationships that are frequently made up of life-giving paradoxes. In short, the possibility exists, not of a universal time, but of the universality of time, i.e. the dynamic interplay of the multiple links between the different attitudes to time.

The idea of one universal time setting a standard for individual times seems on the contrary to be unfounded. For one problem will remain unsolved: the pretended requirement that not only the different sectors of society but also the different societies themselves should be contemporaneous implies that all societies tend towards one single ideal. But nowadays it is frequently said that, in Europe, the Greek ideal of measure and harmony is out of date, as are the Roman ideal of civic grandeur, the ideal of chivalry of the Middle Ages, the seventeenth-century ideal of order and reason, and so on. Do these various ideals follow on in a straight line from one another? Does the ideal, or rather the sense of self-respect despite poverty still deserve a place in societies concentrating on economic matters?

One other point must be mentioned before concluding. In the sociological study of time, is the real question modern versus traditional? Or is this a thing of the past? Does change imply a transition from the attitude to time of 'traditional' societies to that of industrial societies as they exist today?

A better appreciation of the relationship between modern and traditional demands a more precise idea of the criteria for defining modernity. A 'modern' society will have a powerful economy founded on industry, advanced technology, and large-scale capital support, keeping constant watch on new techniques, and on markets to be held or conquered. This logically implies aggressive defence, leading to what Clausewitz, speaking of war, called 'escalation towards the extremes'. Fearing the loss of one of its markets, a 'modern' society may either be drawn into conflict with the other countries involved, or decide to 'protect' any society considered weak because it does not believe the economic all-important. This would seem to show that the ideal of a 'modern' society is an absence of ideals; or rather that the 'balance of terror' is the only kind of ideal that seems to emerge from this competition.

A 'traditional' society is one which does not accept this new 'ideal' of the balance of terror, but believes in values dismissed by the strong on the grounds that they represent an addiction to the past. Such societies are told: until you are as strong as the strongest you will count for nothing on the international chessboard, and you exist only if you count. Until you imitate the strongest, you will simply be pawns. Drop the weight of the past.

Today there is every reason for thinking that the problem posed by the confrontation of the attitudes to time of different cultures is not a question of traditional versus modern, but depends on relationships between forces and the desire for power. Societies are, in fact, always changing. As their attitude to time varies at their different levels—political, religious, ideological, economic—relations between these forces provoke change. Such change is crucial at times of crisis, when relations between these forces break down or are rearranged.

These change-begetting crises can be either exogenous or endogenous. The endogenous crisis is probably the more beneficial to the society in which it occurs, since it marks a moment of beneficial change in the attitude to time specific to the society as a whole, challenging the former play of forces. An exogenous crisis lasts for as long as the pressure lasts. While it frequently entails changes in sectors, and sometimes a veritable mutation in sectors dominated by technology where, to quote Balandier, 'nuclear energy, automation, computer science, and the proliferation of the mass media are usually regarded as triggering off successions of radical changes',[22] it permeates the society only to a limited extent. However important the mutation, it can produce a real new driving force in the society only by linking up with the attitudes that characterize the other forces. Should a technical factor or any other factor of exogenous change be manifestly unacceptable to other social sectors, the tendency will be to refer back to autochthonous values. The ill-informed observer will then speak of the weight of the past and the negative effect of these values on the all-important business of maximizing profits. But the problems cannot be reduced so easily to a kind of modern Manicheism in which anything concerning autochthonous values is bad, while everything that furthers profitability at all costs is good. In any case, the past should not be interpreted solely as a 'weight'. The past may be regarded as a permanent dimension of time in any society, either as a value or as a recourse. In the latter sense, there is no society which does not refer back to the past in times of difficulty.

As regards, for example, the special position in certain parts of

Africa, that man of action and thinker, Kwame Nkrumah, pointed out that the past of a society may be made up of several sedimentary layers which have settled out and are continually being reactivated. He described as 'consciencism' the map in intellectual terms of the disposition of forces which will enable African society to assimilate the Western, Islamic and Euro-Christian elements present in Africa, and so transform them that they can enter into the African personality. The philosophy called consciencism is that which, starting from the present content of the African conscience, indicates how progress will be derived from the conflict that now agitates that conscience.[23]

Accordingly, change is always taking place, perceptibly or otherwise, in all societies. In its accelerated form, which results in mutations, it does not only give rise to the modern versus traditional dispute. All societies modernize themselves (according to their own attitudes)—unless 'modernism' is taken only to mean the rejection of everything but economic profitability, technical efficiency and massive domination. Present change, unforeseeable alike in origin and outcome, raises rather the problem of inventing new societies to meet the combined requirements of the expansion of science, culture and education. An exchange as between equals among societies, each with an attitude to time influencing and being influenced by that of the others, will preclude any domination by the other's time and consequent defensive withdrawals into time past.

Speaking of the time of consciousness, Husserl writes that each present trails a comet's tail—the past—and that the immediate future is always lived as that which, through the still-existing present, the past materializes. This three-dimensional time of consciousness must be seen in conjunction with collective time which, through social activity, is itself also composed of a present which is prolonged by what it remembers and what it anticipates. It is the attractive diversity of the ways in which past and future affect the present that accounts for the variety of attitudes to time in any society.

NOTES

1. Paris, Presses Universitaires de France, 1950.
2. J. T. Desanti, *Phénoménologie et Praxis*, p. 83–4, Paris, Éditions Sociales, 1963.
3. ibid., p. 120.
4. ibid., p. 137.
5. L. Lévy-Bruhl, *La Mentalité Primitive*, p. 446, Paris, F. Alcan, 1922.
6. ibid., p. 456.
7. ibid., p. 458.
8. G. Balandier, *Sens et Puissance*, p. 105, Paris, Presses Universitaires de France, 1971.
9. A. Touraine, *La Société Post-industrielle*, p. 270–1, Paris, Denoël, 1969.

10. ibid., p. 10–11.
11. Halbwachs, *La Mémoire Collective*, p. 116–17.
12. ibid., p. 126.
13. Balandier, op. cit., p. 102.
14. ibid., p. 106.
15. Lévy-Bruhl, op. cit., p. 517–18.
16. G. W. F. Hegel, *Die Vernunft in der Geschichte*, Vol. I of E. Gans (ed.), *Vorlesungen über die Philosophie der Weltgeschichte* [Lectures on the Philosophy of History], 1837.
17. ibid.
18. ibid.
19. W. W. Rostow, *Stages of Economic Growth*, Cambridge, Mass., 1960.
20. Halbwachs, op. cit., p. 101.
21. F. Engels, *The Origin of the Family, Private Property and the State*, Chicago, 1902.
22. Balandier, op. cit., p. 87.
23. Kwame Nkrumah, *Consciencism*, p. 79, London, Panaf, 1970.

BIBLIOGRAPHY

ALTHUSSER, L. *Lire le Capital*. Paris, Maspéro, 1968.
ARON, R. *Les désillusions du progrès, essai sur la dialectique de la modernité*. Paris, Calmann-Lévy, 1969.
BACHELARD, G. *La dialectique de la durée*. Paris, Presses Universitaires de France, 1950.
BALANDIER, G. *Sens et puissance*. Paris, Presses Universitaires de France, 1971.
BASTIDE, R. La causalité externe et la causalité interne dans l'explication sociologique. *Cahiers internationaux de sociologie*, Vol. XXI, 1956.
BRAUDEL, F. *Ecrits sur l'histoire*. Paris, Flammarion, 1969.
DECOUFFLE, A. C. *La prospective*. Paris, Presses Universitaires de France, 1972. (Que sais-je?)
DELEUZE, G.; GUATTARI, F. *L'anti-Oedipe*. Paris, Les Éditions de Minuit, 1972.
DESANTI, J. *Phénoménologie et praxis*. Paris, Éditions Sociales, 1963.
DUMONT, R. *L'Utopie ou la mort!* Paris, Éditions du Seuil, 1973.
EISENSTADT, S. N. *Modernization: protest and change*. Englewood Cliffs, N.J., Prentice-Hall, 1967.
FANON, F. *Les damnés de la terre*. Paris, Maspéro, 1961.
FAVRET, J. Le traditionalisme par excès de modernisme. *Archives européennes de sociologie* (Paris), Vol. VIII, 1967.
FOUCAULT, M. *Histoire de la folie à l'âge classique*. Paris, Plon, 1961.
FREYSSINET, J. *Le concept de sous-développement*. Paris, Mouton, 1966.
GLUCKMAN, M. *Order and rebellion in tribal Africa*. London, Routledge, 1966.
GOTOWSKI, B. Les futuribles à Tarda. *Démocratie nouvelle*, 1968.
GURVITCH, G. *La multiplicité des temps sociaux*. Paris, Cdu & Sedes, 1962. (Que sais-je?)
HALBWACHS, M. *La mémoire collective*. Paris, Presses Universitaires de France, 1950.
HEGEL, G. W. F. *Die Vernunft in der Geschichte*. Vol. I of E. Gans (ed.), *Vorlesungen über die Philosophie der Weltgeschichte* [Lectures on the philosophy of history], 1837.
HETMAN, F. R. *La maîtrise du futur*. Paris, Éditions du Seuil, 1971.
HUNTER, M. *Reaction to conquest*. London, Oxford University Press, 1961.
HUSSERL, E. D. *Cartesian Meditations* (Translated by D. Cairns.) The Hague, 1960.
——. In: M. Heidegger (ed.), *Zur Phänomenologie des inneren Zeitbewusstseins*. Marburg, 1928.

THE PATHOGENIC STRUCTURES OF TIME IN MODERN SOCIETIES

Abel Jeannière

Personal time is always lived as if at a distance from oneself, making it possible to know oneself and to form one's personality. But the individual's time is situated in a far broader evolution: that of human history; it is linked to the external movement of things and of human beings. This linking of inner time with the changing world has always proved difficult for some people. To take two extremes, madmen and geniuses find it hard to bear an outside reality which conditions them in the relationship that they have to establish with themselves. The speed of social change today has made the problem more complicated. The disparity between the rhythm of personal integration and that of social changes resulting from technology may be such that some people lose their grip and can no longer organize their personal evolution in such a way as to understand and live it.

The following pages deal with the emergence of the pathogenic structures of time—pathogenic meaning not necessarily bad, but requiring such changes of rhythm as to make easy transitions more difficult.

The fundamental dislocation between personal evolution and that of the social system is apparent, even before the change in rhythm, in a time division which makes the former's exteriority to the society more pronounced.

Disjointed time

Modern society has a keen sense of the value of time, coupled with the highest regard for punctuality. The importance of a business transaction is measured to a large extent by the time devoted to it. Economic rationality implies a well-ordered and measured distribution of tasks. The calculation of time is reflected in the precision of time-tables and calendars. Whether it be a matter of succession or of concomitance, exactitude requires that reference to the courses of the sun or moon be replaced by clocks and that these be synchronized. Time, thus severed from the biological and cosmic cycles, has become, as it were, exteriorized. It is really no more than a means of measuring our activities, a measurement which lends itself to divisions of all kinds. The order thus established affects people's psychology in a variety of ways. A certain mental concentration on the time factor is liable to turn the measurement of activities into a measurement of performance. With us, living by the clock has become a disease.

ISOLATION OF WORK TIME

The emergence of work time, which is isolated from other human activities and whose duration is a matter of dispute, is a recent phenomenon. At present, work time is measured during the day in terms of a number of separate hours, during the year in terms of alternating holidays and vacations, and during one's whole life in terms of the number of years from the beginning of one's career until retirement.

These time divisions were unknown in the agrarian world. By agrarian world, I mean the civilization embracing highly diversified cultures in which production comes mainly from agriculture and crafts. The whole of time in agrarian life is that of work to rhythm. It has been said that the country-dweller identified himself with his work, though this is perhaps an exaggeration. In any case, work dominated everything. It was omnipresent and everything in the various cycles of the day, of the seasons and of life was organized round it.

The abstraction of time presupposes the abstraction of work: the two go together. Agricultural work is a way of living in complicity with nature, of matching wits with her. The farmer knows his land; the craftsman exploits the qualities in wood and iron and gets round their defects. Work unites them with nature as they transform her. On the other hand, modern work increases the number of intermediaries between man and nature and becomes abstract. Mastery over materials is incompatible with matching wits or complicity. The attention devoted to materials is replaced by attention devoted to machines. And the machines themselves gain their independence in factories that are increasingly computerized. To work means concentrating on various strictly organized signals. Manual skill is replaced by the capacity to respond as quickly as possible to unforeseen stimuli. Complicity with nature, which presupposes imagination and creative ability, is replaced by integration in a code of signals where any sort of initiative is excluded. Simultaneously, there is a transition from concrete, lived time to exterior, abstract time.

Agricultural work was longer or shorter according to the length of the days, the intensity of the sun and the frequency of rainfall. It was regulated by the seasons. Leisure time and holidays were fitted into the year to celebrate the grape-gathering and harvesting periods. Modern work time is a series of isolated hours artificially cut out of the unchanging twenty-four-hour day. These hours add up from day to day, independently of the seasons, and this, incidentally, must initially be considered as a liberation. Henceforth, the variability of work time is linked to the human will and is independent of the cycles formerly imposed by nature.

But, in isolation, work time assumes a value different from other kinds of time. The agricultural labourer and the apprentice craftsman of earlier days were hired by the year. Their hours were not counted; the rhythm of their lives was simply geared to that of their employer. This might mean partial inclusion in the life of the family, as a servant, or it might even mean slavery. In any case, it was the man himself whom the employer hired. Modern employers merely hire a certain number of hours, which therefore acquire a market and social value different from that attributed to other kinds of time.

It is not my intention here to measure the magnitude of this economic and social phenomenon, but to measure that of the reversal of mental outlook. We are led to consider that these isolated hours are the only fruitful ones because they alone bring in money. Their intensity is increasing. Modern business firms systematically organize work time that has been isolated beforehand; they appropriate a number of hours which they must exploit to the maximum for production.

I am not going to refer to the classic theme of the geographical distance between the place of work and the home. The organization of space and that of time are always closely linked. And many and varied are the factors which dissociate work from the rest of life. But I shall draw attention to another kind of distance, of a socio-political nature, one which raises the problem of power.

Work time has thus become a specific number of hours made available to organizers for the purpose of extracting the greatest possible yield from them. The various representatives of the economic powers-that-be will measure the fruitfulness of this time in terms of products. This results in another reversal of values, because the hours are fruitful only in terms of an accumulation of objects which are alien to the very people who have produced them. We are far removed from the farmer who gazed at his freshly ploughed field, or the craftsman who examined and handled his work with pride. The fruitfulness of the hours, too, has become abstract; and so the number of working hours comes to be regarded as time imposed upon us, as a penalty. Increasingly rare are independent workers or, more simply, those who find fulfilment in work which they have chosen, or which they at least like. For the great majority, it is a question of selling their hours, and the buyer will make them as profitable as possible by increasing their yield.

THE 'RATIONAL' ORGANIZATION OF WORK TIME

It is pointless to repeat the history of the organization of labour since Taylor's time. The aim is always the same: to produce more by producing faster. The basis of the Taylor system was the predetermination and control of all tasks. Hence, the best use of a firm's means of production implies the best distribution of operations in terms of time, i.e. their systematic co-ordination. H. L. Gantt, one of Taylor's first collaborators, tried to solve this problem scientifically. More and more research was devoted to planning and the logical flow of work, with the PERT method representing perfection in the field.

PERT (Programme Evaluation and Review Technique) is a method which was developed in 1956 at the time of the creation of the United States nuclear striking force (the *Polaris* Project).

Apart from the technical difficulties, the problem was to ensure the control and co-ordination of the project—to make the weapon operational within a given time-limit and at a reasonable cost by co-ordinating the activities of 250 prime contractors and over 9,000 sub-contractors.

The object was to save time. The importance of this is clear if one realizes that a delay of one week in the manufacture of the gyroscopes meant a delay of three weeks in the adjustment of the guiding system, which in turn meant a delay of two months in the launching of the submarine. The chains of activities and their interrelationship were so many and so complicated that a diagram had to be made out showing the logical structure of the project. And the *Polaris* Project was completed more than two years ahead of schedule.

PERT, and now other methods derived from it, implies an organization of time in which the words 'logical' and 'rational' apply solely to the conditions for attaining the objective. Such planning is undoubtedly necessary and it will clearly have beneficial effects. No contractor would dare to launch a complex public works operation or a large-scale publicity campaign without having recourse to it.

The aim is to make a considerable number of working hours as fruitful as possible by interconnecting them. In short, this means organizing work time without any direct reference to the workers.

This presupposes a precise analysis of the tasks to be accomplished, taking into account the whole complex network of interdependent operations spread over a period of time and directed towards the objective which, in the business world, is always greater productivity at lower cost. A distinction is drawn between operations which can be carried out simultaneously and those which are perforce consecutive, and each task is given an operational duration on the basis of which it is possible to establish a reduced estimate of completion dates. In short, the less time lost in production the more logical the programme will be.

One can only rejoice if this leads to a better administration of many co-operative efforts, but what happens when the 'rationality' of the work thus planned affects everything, down to the succession of motions made by each worker?

According to the logic of the system, no motion should be useless. The interconnexion of motions is such that the slightest pause must be foreseen if it is not to result in multiple delays. Gone is the time when the farm labourer could straighten up for a moment to wipe the sweat from his brow and survey the landscape. Rationalized time creates tension in abstract work which generally calls for mental concentration rather than muscular effort.

And the attention required must be devoted to elementary instructions. Consequently, in so far as the system operates without the worker's being concerned with its logic, his working life becomes unendurable because it is mechanized. Palliatives will have to be found, such as the brief periods of gymnastics which break the monotony of work in certain offices and factories. There is also talk of 'making tasks more meaningful', i.e. the possibility of introducing a little more variety into motions, of acquiring a little more understanding of the task itself and of the whole scheme into which it fits. These are indeed only palliatives, since the aim is always to make the hours more fruitful, more paying although this now implies, taking account of the psychological limits beyond which the worker, considered as a production tool, cannot go.

A more comprehensive conception of 'rationality' and 'logic' to include the 'human factor' must be developed. In the sphere of work, only that which is systematically organized for man's benefit is logical and rational. And no man can be defined and recognized simply by the intensity of his productive work in a system of time economy.

Many factories and administrations have by no means attained such scientific efficiency. Time is therefore subjected to other constraints such as clocking in and hours wasted in pretending to work.

In any event, more and more workers are being driven to regard the most important hours of their lives as outside their working hours. This is going to give rise to the problem of organizing leisure. Once time has become disjointed, problems of integration can be posed only in the form of partial organization, even when it is a question of leisure. Just imagine programming time which, of all things, should be the most spontaneous! But it seems hopeless, at least for the time being, to seek a solution in a better utilization of non-working hours. This is true for two reasons in particular: first, leisure time centres round interests of the same cultural level as those of working hours. A man cannot switch suddenly from deadly dull work to creative leisure activities. Second, the problem is not to take advantage of the separation between work and leisure, but to restore unity by introducing different rhythms.

TIME DISTORTIONS

In speaking of this work area which is isolated in human time, we reason too often as if it were a matter of cutting up a series of uniform hours. Now, to represent the evolution of time by a line or a mere curve is to give a deceptive picture of it. To say the least, it means referring to an abstract resultant which does not allow for the components and breaks. Concrete time, the time lived by an individual or a community, is the resultant of the various rhythms of many different elementary evolutions. This resultant may be to some extent harmonious, but it remains perpetually precarious and is always partial.

The bracketing of hours during the day for intense production gives rise to particularly serious situations. I would mention again the essential distortion already cited, namely the destruction of the harmony established in agrarian society between the rhythms of human life and the natural cycles of days and nights and the changing seasons. That harmony presupposed the constant interfusion of work and leisure. Today the discontinuities in everyday life have more immediate consequences. A man finds himself torn between the speed of his work rhythm at the factory and his need for rest which makes him long for calm periods with no need for hurried motions. This tension between the feverishness of work and the longing for the slow rhythm of rest affects all motions in social life. Many acts necessary to the establishment of human relations will betray the same sense of urgency. In order to reach the calm spot in the country or the peace of home more quickly, a man will race along the motorway, maintaining at the wheel the very speed and degree of tension from which he is trying to escape.

The impact of these new imbalances is combined with the difficulties of earlier distortions. For underlying the recent dislocation of time, old differences in rhythm continue to exist and make adaptation even harder. Generally speaking, technical invention is greatly in advance of political changes. These are never concomitant with changes in mentality, but are the result of power relationships which are symptomatic of disharmony in social development. Lastly, legal justifications and administrative arrangements come after social changes.

The problem of development might itself be analysed in terms of distortions in temporal rhythms, e.g. the disparity between the speeds of development of different economic sectors, the fact that political awareness precedes economic progress, and the co-existence of artificial modern rhythms with natural rhythms made to seem archaic because of the desire for development. In the industrialized countries, this causes the rural areas to resist invasion by urban complexes.

Many people find it difficult to adapt themselves to different though simultaneous speeds of change. It is even more difficult to accept discontinuities and to transcend them personally through an integration which society no longer facilitates. Social life has been hit too hard by the speed at which techniques and relationships are changing. The malaise is due not only to men's desire to escape from a speed which seems to them a pointless rush; maladjustment must also be considered as too great a divergence between the speed of development of the material necessities of modern life and the speed of change of mentalities and psycho-social structures.

Some distortions are probably beneficial, since they create an internal tension which is indispensable to progress. Others, on the contrary, unduly delay necessary developments because people are afraid of them or fail to understand them. In any case, we know that the homeostasis of this world of ours, viewed as a complex system, can be maintained only by conscious political intervention. We have lost the simple faith of the first liberal economists. There is no longer a belief that, by virtue of the natural laws of economics, contradictions will counterbalance one another and the balance will be spontaneously restored.

It nevertheless remains true that, today, each individual, whether by himself or in a small group, must create his own private harmony in the welter of 'uncertain times'.

The acceleration of history

The belief that history is gathering momentum has henceforth become pretty widespread. Since it is more often referred to than analysed, it instils a vague uneasiness, bound up with the conviction that all observable phenomena are now occurring at increased speed. In the end, if this impression were entirely correct, there would be changes every second and the speed of evolution alone would make it impossible to set any course. But the phenomenon is quite different. In point of fact the main difficulty in adjusting ourselves springs from the disharmony between the various rhythms of change rather than from their acceleration.

THE PHENOMENON OF ACCUMULATION

The acceleration of history is evident first of all in an accumulation of phenomena in a number of simple areas which are easy to recognize but difficult to control. The world population growth is an example that need not be dwelt upon here. In certain fields, such as that of engine power or travelling speeds, the increase seems sudden and enormous. Watt's machine in 1785 was rated at only 75 h.p. It took Julius Caesar and Napoleon roughly the same time to cover a distance equivalent to that between Rome and Paris. The doubling of the world population in an ever-shorter time strikes the imagination, but contrary to what Malthus thought, the significant curves of technological progress develop at the same exponential rate.

The time has come when the speed of accumulation is obvious and is bringing about a qualitative change. For thousands of years man's universe was, in his eyes, a stable one. The upheavals of history were regarded as the ebb and flow of a changeless sea. And while time was no longer experienced as something cyclical or even linear, it still moved in a uniform fashion, without any acceleration. Today, man's world evolves at a pace that is measurable in terms of a human life-span. Man's time-scale has changed.

While history is moving faster, the planet is shrinking and man's world seems to be contracting. There is one field in which the combined effects of an accumulation of phenomena and increased speed have a particular psychological impact and that is the field of communication. Whereas yesterday news was usually regional and spread but slowly either as fast as a messenger could convey it or, in some exceptional cases, at the speed of signals transmitted from hilltop to hilltop, all news today can be broadcast throughout the world. It once required a considerable effort to make news known; today, the opposite is true—it takes a great deal of effort to conceal any notable event. Often the only solution is to camouflage it, as it were, by false reports.

THE LAW OF RELAYS

These accumulation phenomena do not spring from a uniformly accelerated growth. Only the abstract resultant may give this impression. Relays exist, as Figure 1, drawn up by François Meyer and showing the increase in speeds, clearly shows.

The motor-car succeeds the horse, and the jet takes over from the propeller-driven aircraft. Each system of transport is perfected and levels off before being replaced. The previous system does not disappear but is used differently. Spacecraft might be added to the list, but do they form a transport system comparable to the previous ones? With rockets we find ourselves in a different situation at the crossroads where engine power and speed meet. As relays increase in number, they change the very nature of the phenomenon.

The components of each acceleration curve must therefore be studied. It is necessary to discover the correlations between the various factors, and to follow the development of each component, both separately and in its relationship with the resultant, which is nothing more than a regrouping based upon a probable evaluation.

The law of relays recurs in all research dealing with a fairly broad section of evolutive time. It is applicable in many fields in different ways.

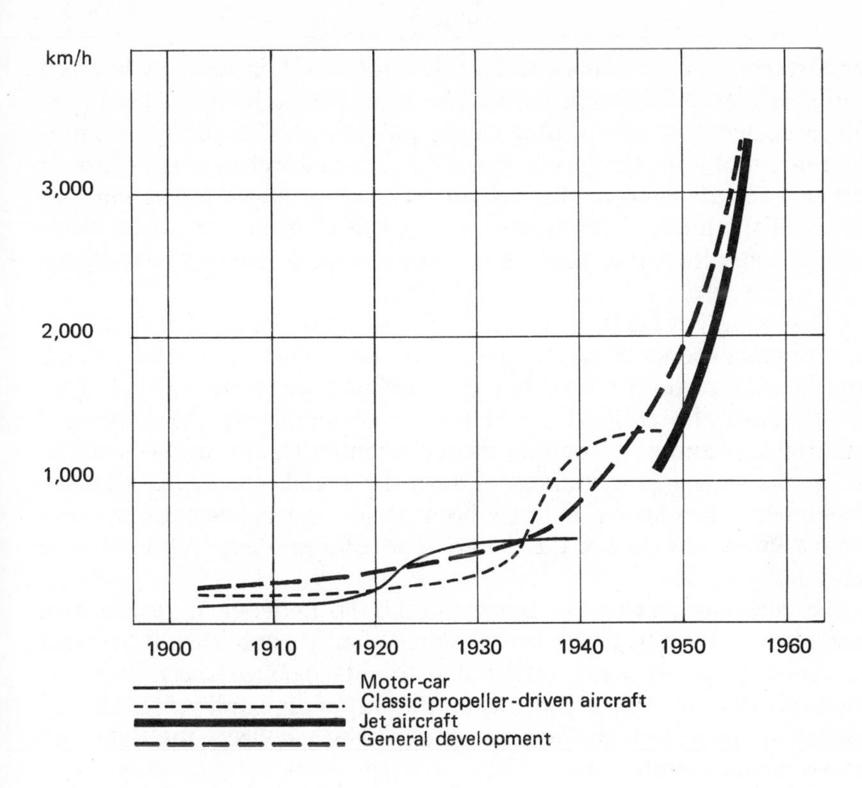

FIG. 1. Increase in speeds.

It is observable on a millenary scale through the changes of civilization. For example, primitive men, as hunters and gatherers, had a very different conception of hunting from that which farmers were to have. For the former, hunting was a harsh, vital necessity, whereas for the latter it was a supplementary leisure-time activity. In neo-industrial civilization a certain type of craftsmanship becomes a luxury and superfluous. And so, what is for one epoch the very nucleus of work slowly turns into a leisure-time activity for the next.

The importance of this law of relays is that it firmly checks the temptation to extrapolate acceleration curves. The extrapolation will be formal and rather pointless unless it remains very limited in scope and in the fragment of the future which it covers. To the problem of divergent rhythms according to objects, places and cultures, the law of relays adds the unknown factor of innovation.

DECLINE IN IMPORTANCE OF REFERENCE TO THE PAST

The future must be approached by more complicated methods than mere extrapolation. The fact remains that the past has, at one fell swoop, lost its importance, not, of course, as a subject for study or curiosity but as a frame of reference for action today. The value formerly attributed to experience gained has been shifted to the possibility of handling an enormous amount of information efficiently. Situations are changing too fast for experience accumulated in the past to serve as guidance for action that must be taken today. This is true at least, and primarily, in all fields involving technology. But technology is not neutral, it is action by technical means, which affects all forms of behaviour.

Cybernetics has taken over from experience. For the head of a business firm, business experience is not enough. It may even be a handicap if the person in charge clings to familiar but outdated work methods. It is far better to have the twofold possibility of accumulating the maximum amount of data and of processing them efficiently. Data must be accumulated on the state of the market, and on the social, economic, financial and technical phenomena at levels both above and below the firm, and one must know how to use the data as part of a strategy that is at once general and specific

The consequences of this are manifold. To understand such a firm becomes more difficult. It will be far more soundly established if founded on its short-term prospects rather than on its balance sheet. But this immediate future no longer depends on the will of an employer; authority is spread anonymously through the group which collects the data and processes and structures them.

Psychological reactions to the decline in importance attached to the past are many and varied. Depending upon a person's temperament, the immediate present and future are perceived either as a crumbling of values or as a liberation. Some see the collapse of a society which they regarded as a stable structure providing a framework for human relations; others view changing morals, ideas and values as something promising.

But with the devaluation of the past, elderly people are being discredited—this is the most deplorable of the consequences. To be old is no longer exceptional and the aged have ceased to be venerable. An old man's experience is but an account of a past with nothing to offer us, and such interest as he may arouse is like that created by old tales which are valueless in modern times; he is no longer held to be a wise man. Unable to assimilate the mass of information required for action, he lives on the fringe of the community. The 'third age' is placed beyond the pale of our societies. The old are a nuisance. Since they are useless, they are absent from the new cities and large architectural complexes; more often than not, they are forgotten in old people's homes where society has nothing better to offer them than slow euthanasia.

The devaluation of the past by the acceleration of history makes some adults feel outstripped by the speed of things before they have even reached the threshold of old age. Their maladjustment is not subjective; they are incapable of co-operating efficiently in the common task. In the most evolutive branches of industry, acquired knowledge soon lags behind the stage reached in technology and a man must be re-phased or 're-trained'.

But it is not enough to offer new knowledge or add cumulatively to that already acquired. A continuous curve gives only an abstract idea of the acceleration of history, whereas renewal is made up of a great many micro-innovations. The law of relays makes qualitative change essential. The psychological manifestation of this novelty comes when new knowledge can no longer be assimilated by the recipient, when it no longer fits in either with his theoretical knowledge or with his mental schemata.

Refresher courses might therefore just as well be called 're-education', since the man who wishes to keep up with the changes in the world must, in certain fields, change his way of thinking and his methods of analysis. The speed of change may also require re-education in human relations because of the changes that have taken place in the conception and style of work in common.

Requirements of this kind weigh heavily on the minds of adults. Some are capable of appreciating the tremendous dynamism of an evolution in which acquired knowledge becomes challengeable more and more quickly. They are glad that at last each generation must invent its own motions and their regulation, at least in a few specific fields in which they are most involved. Others feel an unconquerable unease. This is because traditional acquired knowledge is twofold: there is a technical tradition, that of the material, its use and the human relations conditioned by it, along with a tradition of meaning. Until very recently, the stability of technical material was sufficient to enable children to grow familiar with a world in which continuity was such that they lived in a context of social relations governed by recognized moral values. But the two aspects of tradition are interdependent, and many adults who would find it fairly easy to reconcile themselves to the development and transformation of technical material, are at a loss when they realize that material innovations have destroyed accepted values. By definition, new values do not bear the stamp of approval of the past which seems to them an essential guarantee. But innovation changes the balance of all the traditional structures and furthermore calls for the discovery of meaning.

The escapist solution will often be to get bogged down in the present of the so-called 'consumer society'. This implies giving up the struggle, drifting with the current, or clinging to a job, to a salary, and holding on until retirement, a retirement which comes early enough to allow one to get about still and is sufficiently comfortable to minimize the risk of being forsaken in old age. The over-pressing weight of the immediate future, the acceleration of change and the multiplication of distortions condemns many an adult to the moroseness of a barren present.

What about young people? The education required by the acceleration of history ought to place less emphasis on the transmission of a heritage of knowledge and more on the capacity for innovation and adjustment and a taste for constantly renewed study. Education must have a dynamic content, which is difficult to define since it is still in the planning stage.

The time of youth

Youth is prized more today since the short- and medium-term future conditions human activity because the acceleration of history calls for detachment from former modes of action. The more dynamic part of society associates youth with its own taste for initiative. But young people find themselves projected, without any experience, into uncertain times in which the arabesque of distortions is unpredictable. The phenomena analysed in the first two parts of this essay have particular repercussions on young people who find themselves at once more highly valued and frustrated. This sheds light on the 'generation gap' which deserves a lengthy study.

INTERMEDIATE TIME

Grouping young people together in order to study them as a specific group is a recent result of the social trend. In traditional civilizations, the transition from childhood to adult life takes place without the intrusion of intermediate time. The break is clear-cut: the little girl becomes a woman when she is capable of bearing children, and the boy is an adult when he begins to work at, or at least to learn, a trade. The transition is usually sacralized by some sort of a religious initiation. In industrialized societies, an intermediate age develops during which young people, without being allowed to assume civic responsibilities and without sexual segregation, prolong their preparation for adult life—a preparation which excludes them from taking decisions in the civic and political world.

This is the new age group which is going to find its value enhanced as the source of potentialities for the future, and which may influence the development of present structures. Unfortunately, this enhanced value is based only on negative, vague factors. Negative, because young people are not hidebound by customs and institutions; vague, because among their number are those who will play an effective part in history, though which of them or how, we do not know. In point of fact, young people as a group are only helping to enhance the value of the immediate future. For it is too often forgotten that the weight of the past is not defined in its entirety when it is treated as a handicap. Certain rhythms remain slow in contrast with the acceleration of others and ensure some essential continuity. Two processes remain unavoidably slow even in our time—that of the maturing of the individual and that of the cohesion of an expanded group.

MARGINAL TIME

We value those who will judge the rightness or wrongness of our predictions. In reality, young people are a provisional exteriorization of our short-term future. They live on the fringe. They are not among us. They are in a hostile space designed neither for them nor by them. We prepare them with our adult outlook for a world which we are building in a hit-or-miss way. Education, as it is still conceived in most secondary schools, draws on old ideas worked out at a time when social movements were slow and when spontaneous contact with the natural environment sufficed to ensure the harmonious development of young people. But children cannot extend too far into adolescence the stage of games in which they create their own symbolic world. The intermediate age offers no transition to the stage where the young can take action to transform the surrounding world. It is a time apart, a marginal time.

With young people, work time and leisure time are also separate. Neither fulfils the conditions for adequate socialization, or even for anything but artificial inclusion in the community. In their enclosed space, young people's activity is subject to merely simulated regulation. They live in marginal time which is cut off from real modern life. Leisure and work follow the hectic rhythm of a strict and over-loaded time-table; it is only a partial imitation of adult time.

DIFFICULT LINK-UP WITH THE FUTURE

It is a matter here of linking marginal time to the over-all time of human evolution.

Less and less taken in by the enhanced value conceded to their age group, young people are induced to measure the time imposed on them for acquiring the necessary competence before entering a society based on the intensive twofold utilization of energy and information. Marginal time fosters spontaneous creation, but one without a future. Entry into society reveals the rigidity of the present. Probably, in marginal time, links with the past seem like chains which have finally been shaken off. But it is permissible also to assess the magnitude of the necessary task of mastering the flow of changes by proper handling of information in any field whatsoever.

The lack of coherent connexion between a shattered present and an uncertain future rules out truly creative activity, which is, at the same time, an assertion of oneself. Now, as long as it is impossible to create, or as long as the individual is obliged to create in a vacuum on the temporal fringe of the social movement, he is naturally seized with a strong desire to destroy. The violent conduct of a number of young people is in fact an unconscious protest against the absence of integrated time, against the marginal nature of their time and that of the social space in which they are confined.

How is it possible to break out from a temporal dimension which runs parallel to the evolution of society but has no communication with it other than a verbal one? The possibilities are very limited.

Some young people can play the game, be 'good students', accept confinement in studies without understanding their usefulness, and prepare with apparent calm for the diploma or other form of final degree which will put an end to their marginal life. But this prospect is denied to the majority of young people, for whom the time of studies and apprenticeship leads only to a disjointed time from which will be subtracted the dreary hours of a dull job. Society can expect nothing more of them than an increasingly unreliable resignation.

A link with the immediate future being impossible, an alternative course is flight to Utopia. Young people's Utopia is not the logical creation of an enclosed world; it has nothing in common with Thomas More's island; it does not lie, unconnected with us, in the ocean of the future. It consists of Utopian trends aiming at the achievement, in an idealized abstraction, of some of their most vital aspirations. What society can we imagine that would foster spontaneity in both working life and sexual life? In a word, a future which reverses the values of present-day society is projected in semi-tones, at once imagined, desired and half-experienced. In contrast to industrial efficiency which adjusts men to the rhythm of mechanized production and connects up combined series of machines and men, we have the creation of a community whose primary concern will be harmony in human relationships. In contrast to the 'rationalization' of time with a view to maximum output, we have the rejection of any standards in the use of human resources. In contrast to school curricula geared to entry into industrial society, we have educational content for its own sake. In contrast to industrial artificiality, we have nature, but not nature as it has become, tamed and just as artificial as any other industrial object.

This dream-future, stemming from reaction to the harsh realities of today, may be realized in two ways. One is to create small communities which, living on the fringe of society, will achieve some of the future harmonies. But the paradox is that to escape from an enforced marginal world, another such world is created intentionally by a deviant group, and seen as the incarnation of new values. Some may wish to fight effectively to bring about this dream-future and will transform their protest into a political struggle. But this struggle, too, will be marginal, because young people will find it difficult to identify their demands with the present-day orientations of the big traditional revolutionary parties. A revolutionary party does not rely on fringe people, even if they are young; it is linked to the huge working class. And so, in order to come out of the ghetto and get back into the thick of the social struggle, there is no alternative but to go directly to the workers. But the workers, in their turn, are suspicious of these well-meaning visitors. To be a worker is not a calling.

Another solution is to flee the marginal present for the present itself. This is pure escapism—not that of the adult who settles down in the consumer society, but an escapism in which the present is regarded as an evasion. This means that every second will be lived with the maximum emotional intensity, by means either of drugs or sex. But there may also be a certain understanding of art. Or there may be recourse, not to integration in a group, but to collective demonstrations which merely agglomerate separate individuals. Having no hold over society, the easiest way is taken, that is, to speak out against it.

The prospective approach

What must be found is an approach that is at once modern and felicitous, one that is an effective antidote.

There is no partial answer to these sicknesses of lived time. The pathogenic structures of modern times can be modified, but only as structures. Acceleration and distortions cannot be eliminated. What must be contemplated is both a radical change in mentality and a form of action that is collectively effective.

The change in mentality is absolutely necessary for anyone who wishes, first of all to survive, and then to thrive, in the coming world. The remedy exists and we know what the new attitude is—it is that of the prospective approach. The aim of prospective research is to enable the individual to relax in the accelerated rhythm of changes and to allow him to act with the future in view. The prospective approach is an attitude of mind deriving from observation of two phenomena: the destruction of an earlier situation in which man could easily find his place and understand himself, and the absence of any clear historical determination of the future by the past.

Prospective research is something quite different from extrapolation or forecasting.

Extrapolation is the extension of a movement whose rhythm and acceleration it has been possible to analyse. It is the continuation of past time. In extrapolation, we fancy that we are imagining the future, whereas we are in fact only tracing a curve in time that is past.

Forecasting has a wider framework. It presupposes, in the present, a system of relations that can be correctly analysed. It then becomes possible to vary or foresee variations in certain terms or tensions by measuring the reciprocal importance of their movements. It is even possible to introduce one or more new variables which create a new balance in the whole. Forecasting is always concerned with the evolution of a known system or coming modifications in a system whose mobility is predetermined, probably only in part, but enough so as to make an explicit statement on the future possible.

But forecasting becomes impossible when it ceases to be restricted, because evolution is a ramified process and no one knows what branches may survive or which of them will be the most flourishing; because the future will be the fruit of a number of acts which will have to be chosen and carried out; and because the decisions to be taken lie at a crossroads of projects which do not form a system, but, at best, series of systems which must not be unified if we are not to imagine present reality instead of analysing it. Prospective research faces up to this uncertain future; it deals with the effects of present action on the future; it leads to experimentation, whereas forecasting can lead only to verification. As regards the present, prospective research has a few certainties to go on; for the future, it offers only hypotheses and strategies.

The prospective approach is the general attitude required today for action. It sees today's actions in the light of a number of possible futures. It represents a permanent desire to choose, from among present means and objectives, with an eye to the future. Possible objectives which we wish to attain must be substituted for events which seem to be part of the nature of things, events which are imposed on us and must be endured. It is not a matter of divining or describing the future, or indeed of reckoning what almost inevitably awaits us. What it means is to determine, in the midst of present objective necessities and their carefully measured evolutive weight, the means of making history which are still available to us.

This probably brings us back to the calculation of probabilities and the games theory. Our situation is that of a player who must invent the rules of a game in which the figures are always new and are never repeated from one game to another. Harmonies and discords of lively but slow-maturing rhythmic combinations, ramified possibilities, spontaneous apparitions, fleeting initiatives and prodigious orchestrations belonging to no over-all system sum up the maze of relations established and broken at many different levels. The prospective approach here shows its fruitfulness which consists of maintaining the joy of research while remaining consciously in the realm of uncertainty and hypothesis. The means of fighting against the pathogenic structures of time are to be found not in things, but in men.

Applied to projects of some magnitude, prospective research relies upon the calculation of decisions. We are not without weapons to tackle the multiplicity of probabilities. It is possible to work out a strategy if we remember that, in the economic and social fields in particular, all the options must be coherent, that the interconnexion of results must be foreseen, and that, furthermore, the progression of causes and effects must remain flexible and workable.

Without examining the political consequences of the prospective approach, we know that a radical change involving nothing less than a complete transformation of values is required. The structures of time cannot change unless man himself changes. The disintegration of meaning, as handed down like a heritage from the past, does not only throw us back to uncertain times but also commits us to a political effort to create meaning. Such a change cannot be worked without leaving on the fringe those who are made ill by changes of time. Here again it is a political effort and the application of prospective research that will make curative, logical and concerted action possible. The future will be, in large measure, what we make of it.

BIBLIOGRAPHY

Antoine, P.; Jeannière, A. *Espace mobile et temps incertains*. Paris, Aubier-Montaigne, 1970.

Berger, G. *Phénoménologie du temps et prospective*. Paris, Presses Universitaires de France, 1964.

Étapes de la prospective (collective work). Paris, Presses Universitaires de France, 1967.

Federal Electric Corporation. *The P.E.R.T. method.*

Grossin, W. *Le travail et le temps*. Paris, Éditions Anthropos, 1969.

Hetman, F. *The language of forecasting*. Paris, SÉDÉIS, 1969.

Jouvenel, B. de. Le langage des heures. *Analyse et prévision* (SÉDÉIS), no. 4, April 1972, p. 437–70.

Lakomy, Z.; Novy, O ; Gottlieb, M.; Czerny, M. Cadre de vie et civilisation. French translation by Bernard Cazes in *Analyse et Prévision*, SÉDÉIS, no. 4, October 1968, p. 631–44.

Leclère, R. *Les méthodes d'organisation et d''engineering'*. Paris, Presses Universitaires de France, 1968.

Meyer, F. *Problématique de l'évolution*. Paris, Presses Universitaires de France, 1964.

Paz, O. *El arco y la lira*, 1956.

Rambaud, P. *Société rurale et urbanisation*. Paris, Éditions du Seuil, 1969.

Psychological and moral valuation of time

THE PHILOSOPHIC CONCEPT OF TIME

Yakov F. Askin

Time, like other problems, began by being the exclusive concern of the philosophers but, as separate sciences emerged, it attracted the attention of scientists as well. But the study of time remains one of the most important tasks of philosophy. The philosopher accepts this task not simply out of tradition but because it is a vital part of any philosophical attempt to understand the essential features of the changing world to which we ourselves belong and which is reflected in the body of contemporary knowledge.

In addition to occupying an important place in contemporary art and science, the category of time represents the point at which the attention of a number of philosophical traditions converges. In this short essay, we shall describe some of the positions which characterize Marxist thinking on this problem, and, in our opinion, accord with the most recent scientific theories.

The full significance of time emerges only in connexion with the apprehension of objects seen as in the process of change. We speak of time when we refer to a sequence of events or to their duration. But time does not only denote the flow of impressions and ideas in the mental life of man—above all it denotes a mode of existence of the external environment, the dynamics of development of the material world. V. I. Lenin wrote in this connexion: 'Recognizing the existence of objective reality, i.e. matter in motion, independently of our mind, materialism must also inevitably recognize the objective reality of time and space ...' [1]

The objective nature of time is now generally recognized in contemporary science. Physiology, for example, reveals that the orientation in time which is a feature of all living things is a reflection of the temporal properties of the external environment. An important contribution in this field is made by the analysis of conditioned reflexes as a function of time and the study of 'biological clocks'. The latter term refers to the ability of living organisms—both plants and animals—to measure time, regulating their lives according to daily, seasonal or yearly rhythmic patterns. Observation and experimentation have shown that rhythmic processes in living organisms are determined by rhythmic changes in the physical environment.[2]

Some mention must also be made of geochronology. We are often prevented from thinking of time as a fundamental, internal characteristic of all processes by man-made clockwork mechanisms, which to the ordinary view seems to be the personification of time. This can also be seen in the paintings of Salvador Dali, where time is represented by clock faces. To such a view time frequently appears as something external to phenomena. In relating time to processes taking place within geological material, geochronology uses as a clock the process of radioactive conversion of chemical elements and demonstrates that every natural phenomenon, including those of the inanimate environment, is also a manifestation of time and records its passing.

Of course, the objective time involved in processes of development should not be considered as exactly equivalent to time as registered by the observer.[3] Time as experienced psychologically in a physical, biological and social context is one thing, and the rhythmic patterns exhibited at various levels, for example in the microcosm and the macrocosm, are another, which does not, however, prevent recognition of 'objective time, the structure of which is independent of the observer'.[4]

This objective time is not a separate, independent substance but a form of existence. Contemporary science makes it impossible to agree with the Bergsonian philosophy which, in the words of Maritain, sees time as 'the absolute, the source of inventiveness and creativity'.[5] The concept of time as an independent substance has many supporters, from the Orphic religion of ancient Greece, which considered Chronos as having a separate existence from the beginning of the world and as giving birth to the elements of fire, air and water, to Spengler who identifies time with fate, and Heidegger who, describing 'temporality' as a primary form of time, distinguishes it from subjective time as one of the basic categories of existence. 'Temporality', writes Heidegger, 'is the original "outside-self" existing in itself and for itself'.[6] But contemporary science, and here special reference must be made to Einstein's theory of relativity, attests to the reality of a view of time which considers its essence to be a system of relationships between events. This view sees time as a reality inasmuch as it is a form of being, but not in that it is a separate 'thing' in its own right or an independent process acting as a demiurge of reality. Here one is bound to recall Leibniz's perceptive understanding of the essence of time which, as we now realize, was closer to the truth in this respect than Newton's 'absolute time'.

On the other hand there is the view expressed by the Australian philosopher Smart that the concept of the passage of time is an illusion.[7]

A similar position has been adopted by the American philosopher Black.[8] It is difficult to agree with these authors. They assimilate the passage of time to the movement of individual things, with characteristics—such as units for measuring the speed of time's passing—that imply a relationship between spatial changes and time, which is meaningless in this context. And as time is not an individual 'thing', they declare that the concept of the passage of time is a 'figure of speech', a fiction.

But the concept of the passage of time apparently has a real meaning and refers to the changing sequence of events from the point of view of their existence. The world has its history: all its phenomena do not exist concurrently, and the existence of each separate 'thing' has a beginning and an end. This is what is meant by sequence in time and duration in time. Neither the position of Kant, who wrote 'Time itself ... already contains relations of succession',[9] nor the interpretation of Bergson, who in a description of time speaks of 'multiplicity without division and succession without separation',[10] provides a clue to the real nature of the passage of time. 'Bergsonian duration is a duration which does not last',[11] notes the French philosopher J. Pucelle, and we can only agree with him.

Only when we understand sequence in time as the expression of objective changes in nature and human history do we have a scientific philosophical basis for investigating the passage of time and its properties. The passage of time as a changing sequence of events in the stream of existence is related to a certain temporal structure, and to certain relationships between moments in time. These relationships can be defined both quantitatively, in accordance with before/after concepts, and qualitatively, distinguishing one moment in time from another according to its relationship to the process of becoming, distinguishing in other words between what exists and what has existed, what exists and what will exist. It is at this point that the past, present and future, those characteristics of the different 'aspects' of time, appear before us.

In our century, whose intellectual awareness has been coloured in many respects by the contributions of natural science, one of the most important being the theory of relativity, the concept of time is being increasingly linked with the concept of space. In the formulae of Einstein's theory time appears as the fourth co-ordinate of space. This should not, however, be taken too literally. Multi-dimensional space is above all an algebraic concept and physical space is still three-dimensional. Even in the century of the theory of relativity time retains its own specific nature. A close connexion has been found to exist between time and space, but it is only a connexion—the two are not identical.

Before/after concepts can in some respects be compared with the spatial relationships, behind/in front or left/right, but there is no spatial equivalent of such temporal concepts as past, present and future. This is where the essence of time emerges quite clearly. Time appears as a mode of being of everything which exists, a mode which is linked with the process of its development, with the transition from possibility to actuality and with what is defined by the philosophical term 'Becoming'.

Orientation towards the future and recollection of the past are characteristics of human consciousness, and in this sense we do not exist in time simply, as in the present moment, the 'now'. Of course the present is the central aspect of time. It is their relationship to the present which characterizes the two other aspects of time: the past is what was once the present and the future is what will be the present. The present is always correlated with some existent situation or event, the duration of which determines the 'dimensions' of the present, in a manner varying from one category of events to another. The range is extremely wide: from the geological present of our planet (the Cenozoic age covering some 70 million years), or the biological present of man, dated from the appearance of Cro-Magnon man (approximately 50,000 years ago),[12] to the present as a specific state characterizing the conversion of an elementary particle, equal to an infinitesimal fraction of a second, or the present of psychophysiological perception, the duration of which is determined by the capabilities of the sensory organs.

The relativity of the present is also emphasized in our use of language. The Danish linguist O. Jespersen points out that in practice the term 'now' describes a time span which can vary greatly according to circumstances and in some cases be quite long.[13] What is certain, however, is that the duration of the present cannot be zero, for this would mean that the object (event) to which it is related and by which it is determined does not exist. This approach to the present differs from its presentation by the Aristotelian tradition as a specific point in the sequence of time separating the past from the future, resembling what the Thomist philosopher Nys defined as 'simply a boundary or demarcation line drawn by the human mind in the time continuum'.[14]

The present is the expression of a relative stability in the constant process of change. This relative stability is not merely an abstraction; it has an objective basis in the discrete nature of the material world.

The constant motion of matter implies that the present, future and past are linked. This link is essential to the process of development and would be inconceivable without it. Memory is organically linked with time, in fact it is impossible to solve the problem of memory without solving the problem of time.[15] But memory is not the only custodian of the various stages of time. The past is also visible as materialized in the deposits of the different geological epochs, in tools which are the tangible result of the skill and experience of many generations, and in the ganglia which form in the nervous system. The time that a given phenomenon has lasted does not disappear without trace, it can be said to leave its mark in it. We are all familiar with the 'rings' in a tree trunk which show its age. Everything in the world—from molecules and stones to living beings and social structures—has similar 'rings' of its own.

The principle of historicism, which postulates a durable link between the present and the past, is one of the main foundations for scientific thought and practical activity and an essential characteristic of the dialectical approach to the world. Forecasting the future has also become a very popular subject in recent years. Much has been written about the future development of the economy, of science, of the arts, of human society in general. This raises several questions for a philosopher, one of them, and perhaps the most important, being: is this orientation towards the future a purely human phenomenon or is it only a human modification of some fundamental characteristic of the whole process of development?

Philosophy, by generalizing the results of the most recent scientific research, can, in our opinion, reply affirmatively to the second part of this question.

But we should first consider how we are to define the term 'future'. If the past is realized potentiality and the present the transition from possibility to actuality, what is the future? Is it a fiction, an existential vacuum, the product of pure consciousness, or should it be considered simply as something which has already been materialized and exists somewhere, like an undiscovered galaxy unknown not only to cosmonauts but even to astronomers, but which none the less exists?

The future is neither of these things. It is neither a fiction nor simply something which lies beyond human consciousness. The pattern of its development is rooted in the objective laws of reality, it is the realm of unfulfilled but real potentialities. As such it is not, on the other hand, something which already exists, like a point in space which one has not yet reached. To move into the future is not at all the same as to travel to a constellation which already exists, but has not been reached. To move into the future is to create the future. Movement into the future is the process of its creation or realization. And this is not only true for mankind. The world exists in time not merely because time is the medium in which occur such changes as consist in the reshuffling or reordering of what already exists but because time is the crucible in which the new is born and the old passes away. The world does not only exist in the present. It is continually being created—or rather creating itself—in a constant process of becoming, of what might be called the 'accretion' of existing reality, and this it is which expresses the passage of time.

The longer mankind lives and the longer it labours, investigates and reflects, the more clearly a scientific understanding emerges of our kinship with the world as a whole. This is not the instinctive feeling of oneness with the Cosmos which characterized the ancients. In the discovery that what appeared to be a distinctive feature of mankind is just one manifestation, albeit the most advanced, of a general attribute of matter, all the specific qualities of these manifestations are taken into account. Yet there is clear evidence of this kinship, particularly as far as the category of future is concerned.

The contemporary study of physiology clarifies the role of the future in living organisms. Research into the processes of anticipation has produced some interesting results in this sphere. Professor L. V. Krushinsky of Moscow University has identified anticipatory reflexes called extrapolatory reflexes which he has examined in birds. Extrapolatory reflexes cannot be reduced to conditioned reflexes. But even if one considers conditioned reflexes, the Soviet physiologist P. K. Anokhin has explained the classic Pavlovian reflex in terms of its role as a signal, giving a message about the future. In point of fact, the dog hears the footsteps of an attendant bringing its food and begins to secrete saliva which is no use to it now but will be in the future when it has the food in its mouth.

This interpretation does not deny that psychic life is a reflection of objective reality but it is, in Anokhin's words, an anticipatory reflection.[16] This reflection at human level is none other than a manifestation of the general characteristic of reflection, an attribute of matter which is rooted in its very nature, as V. I. Lenin pointed out. What is now suggested is that this anticipatory character is in turn a general property of all reflection.

The study of cybernetics has given a new impetus to rational analogies between human behaviour and that of any self-organizing system. It is no coincidence that the work of one of the pioneers of cybernetics, the Soviet physiologist N. A. Bernstein, shows that the future as a reflection of trends in objective reality and as an expression of the needs of the organism is an important, and to some extent decisive, factor in the organism's development and activity. The idea of modelling the future, of the future as something necessary to determine the behaviour of the organism, plays a dominant role in Bernstein's physiology of activity.[17]

For centuries the future has been the theme of idealist philosophy and religion—both in the pictures drawn by Biblical prophets and in treatises devoted to teleology, describing finality as a spiritual force governing the world. Orientation towards the future, stripped of its mystical connotations, emerges as a basic characteristic of all development. In inorganic matter it is expressed in the irreversibility of a number of fundamental natural laws, of which we need only refer to the laws of thermodynamics. This orientation of development not only does not contradict the idea of causality, but is linked to it and can be seen, if one wishes, as its projection into the future. In fact the nature of any event is determined not only by what has gone before (its causes), but also by the direction in which it is moving, by the tendencies which are inherent in the object that is developing. Forecasting the future is therefore important not only if we wish to know what is going to happen in the future, but also if we wish to understand what is happening in the present. It is possible to understand a given situation fully not only if its past origins are taken into consideration, but also on the basis of knowledge of its future, when the possibilities which have ripened within it will have been realized, thus revealing its essential nature. As Marx said: 'The anatomy of man is the key to the anatomy of the monkey.'[18]

As a general rule we understand the past better than the present, but this is not because we know more about its characteristics than when it was the present (sometimes this is true, but sometimes the reverse is true: much may have been forgotten). We understand the past better than the present because we know what it led to: we recognize the tendencies within it which have ripened, the movement which was hidden earlier and the branches which bore fruit.

This knowledge is 'produced' in the course of the development of the object.

This does not involve any rejection of the idea of the uni-directionality of 'time's arrow', which is linked with the irreversibility of the process of development. The concept of the future is inseparable from the idea of irreversibility: a chain of events in which the events themselves are repeated absolutely and their course is reversible removes any real difference between past, present and future. Only a subjective distinction between them can remain—related to the act of perceiving the events or experiencing them. But a world which is absolutely repetitive and absolutely reversible would itself lack such a distinction: what is to be has already been, and what is has already been and will be again, all potentialities have been realized—things and events would not come into being in the proper meaning of the term, they would merely repeat themselves. The concept of development could not exist in such a world. It is identical with Nietzsche's universe and its 'eternal recurrence'. It is the sort of human world constructed by Freud's psycho-analytical theory, with its basic assumption that the adult is formed solely of his past experiences, that human behaviour is uniquely 'programmed' by the sexual impressions of early childhood which are stored in the subconscious. In this way Freud presents man with his past in the guise of the future.

A world without development is a world without future, without the passage of time, in fact without time itself, for time is the form taken by the process of development. Time without irreversibility would resemble space, as has been clearly demonstrated by the American writer Kurt Vonnegut Jr in his most original novel *Slaughterhouse 5 or the Children's Crusade*. The hero of this novel, Billy Pilgrim, has a concept of time which, so it seems to him, he learnt from the inhabitants of a planet called Tralfamadore and where time resembles space. Everything exists all the time, all events happen in the present, everything is repeatable, including death, return to the past and passing through any given period of time.

But if we leave the fictional planet of Tralfamadore and return to the real universe time does not resemble space, for here we find causes in operation, bringing consequences in their wake, events follow on from each other and their sequence is irreversible. Such phenomena of repetition and reversibility as occur are strictly relative in character. Rhythmicity does not affect the uni-directional, irreversible passage of time.

The determinism of the process of development involves certain restrictions which are imposed on the nature of this process by possibilities. But possibilities, in their turn, might be seen as representing the future, in other words as extrapolations from the laws governing the universe, extrapolations that are implicit in it though not yet realized. The possibilities for the development of matter as a whole are infinite, inexhaustible, but every separate, concrete process of development has its own particular possibilities. On the one hand, the possibilities here are many and various, in fact the many-faceted nature of possibility, the creative nature of development, finds its clearest expression in the essentially statistical character of the basic laws—whether of quantum mechanics, mutations in the course of organic evolution or economic development. On the other hand, in every concrete situation there is a limited number of real possibilities and there are limits to the diversity of the future. In particular, it is a long time since scientists refuted the naïve Lamarckian theories of the unlimited lability of living organisms.

The directionality of the process of development cannot be understood in the fatalist sense of predetermination of the future, for this would make becoming, and development with it, a fiction, and would deprive the passage of time of its real meaning. But the reasonable effectiveness of forecasting techniques in a considerable number of scientific and practical fields bears witness to the existence of a fixed framework, within which each specific development takes place. The presence of a fixed directionality in development is the objective basis for scientific forecasting.

In our century the problem of the future is more than ever present in the minds of men and in their feelings. And this tendency is the expression of a quality which does not pertain to human consciousness alone, but in a sense informs the whole process of development in the material world.

Orientation towards the future permeates man's entire psyche. This is proved conclusively by, for instance, the research into the psychology of attitude begun by the Soviet scholar D. N. Uznadze and now being continued in the Institute of Psychology of the Georgian Academy of Sciences.[19] It shows awareness of the future to be one factor in raising the level of social behaviour. This is true of society as a whole, when it is able to base its actions firmly on scientific forecasting (in the social and, more particularly, economic domains), and is proving to be an essential feature of social management. It is also true of individuals. Orientation towards the future is a character trait typical of the more mature type of personality and a basic pre-condition for a developed sense of values.

The problem of safeguarding the natural environment, which has become particularly acute today, reminds humanity very forcefully that it cannot sacrifice the future to the momentary enjoyment of the present, for today's future will inevitably become tomorrow's present and its account will have to be paid. People must shoulder their responsibilities vis-à-vis the march of time.

The future is linked with perhaps one of the most valuable properties of the human personality and human communities—the property of activity. Awareness of the passage of time is a recognition of the distance between cup and lip, in the words of Guyau, the French philosopher and psychologist of the last century. This is a good description of man's situation in time, a situation which is linked with awareness of the future. Scientific progress has enabled us to see that nature exists on a wider time-scale than we had thought. Social progress broadens the horizons of human time.

Without lingering on other aspects of the philosophic concept of time,[20] we shall now turn our attention to one of the basic properties of time, namely its infinite nature, expressing the continuing existence of the material world, and the fact that its movement has no beginning and no end. The infinity of time is denoted by the concept of eternity. But before this statement can be accepted we must explain in what way the concept of eternity reflects the characteristics of time.

At first sight the definition of eternity as endless time characterizes it merely as a quantitative measure of time, linked solely with the duration. But this view conflicts with the fact that all objects and phenomena existing in the real world are finite, transient, do not have infinite duration.

The attempt to refer to the idea of the eternal as a separate 'thing' inevitably leads to rejection of the material object as a bearer of eternity. Sometimes the fact of the finiteness of things and phenomena, of their coming into being and passing away, is used as evidence that eternity is no more than an idea. Thus the French philosopher Alquié writes: 'Our own frailty generally leads us to suppose that Nature is eternal. But it would be useless to seek a truly eternal object in the given world ... Our everyday experience teaches us that all phenomena take place in time, and the eternity of the object can only be postulated as external to the given world, transcending appearance. It cannot be understood as a thing.'[21] And Alquié goes on to seek for the foundations on which rests the concept of the transcendental subject as a bearer of eternity.

The truth of the matter is, however, that eternity as endless duration is not achieved by the endless 'freezing' of things and phenomena in one unchanging state, but their change and succession. Eternity is achieved, if you like, by the world's undiminished ability eternally to change and renew itself. It is not right to contrast eternity as something immobile, as motionless duration, with the passage of time. There can be no such thing as motionless duration, just as there can be no such thing as roast ice or a square circle. For the essence of duration is the motion of matter: it cannot be motionless. Inherent in eternity are two basic features of time: duration and sequence.

The constant contrasting of time and eternity in idealist philosophy is largely due to the fact that eternity in idealist philosophy is considered to be a state in which sequence does not exist and everything occurs simultaneously. Thus the English philosopher Stocks affirmed that eternal life, existing outside time, was devoid of all distinction between past, present and future, and wrote of the higher order of existence, excluding events and sequence, which represented eternity.[22]

The supposed conflict between 'time' and 'eternity' is indeed most evident when time as involving change is opposed in this way to eternity as involving changelessness. But in our opinion what has been said above shows that there is no justification for this opposition.

To set eternity against time is to set one aspect of time against another, splitting one from the other to make each an absolute, making it epistemologically possible to interpret them in a theological manner. Lenin, following Feuerbach, noted: 'Time outside temporal things = God.'[23] One aspect of time, duration, which expresses the idea of continuity, separated from its other aspect, sequence or change, is converted, in the idealist interpretation of the category of eternity, into an attribute of some kind of unchanging being, an attribute of God.

Eternity is not the absence of change or of activity but rather the unending motion of nature, constituting the form of being of the material world. Nor will continuity, the continuation of existence, be found in existence detached from time but in temporal existence itself, for this continuity or duration in fact derives from the changing sequence of states or of events. The eternal does not exist outside time, outside the real world, whose development constitutes the content of time's passage, but within it.

Eternity is the realization of endlessness and at the same time the continual realization of the inexhaustible creative potentialities of matter, of the unending dialectic of the transition from possibility to actuality.

If one takes the view that eternity is being outside time, that, in the words of the British philosopher McTaggart, 'eternity is the end of time',[24] then eternity is not a positive but a purely negative concept. It would be merely the negation of time and all its properties. As Alquié says: 'The idea of eternity proceeds from a psychic attitude which rejects becoming. It results from the rejection of time.'[25]

But eternity is not something negative: it is profoundly positive. Eternity as the permanence of existence reveals the creativity of the process of time. Time is sometimes identified with mortality, extinction, destruction. Characteristic of this approach is the position of the Existentialists who see time as devouring existence. 'Time', writes Sartre, 'separates me from myself, from what I was, from what I want to do, from things and from people.'[26]

But time does not only represent destruction, it also represents birth, genesis, the transition from non-being to being. Belief in progress implies acceptance of the passage of time.

The purely temporal nature of human life is a basic fact which must obviously be borne in mind by any system of ethics which has as its object live man and not a rough model of man which is completely divorced from reality.

The question of man's mortality has been considered by theologians and by philosophers of the idealist school, but the answer which they give is speculative. The basic reason for this is that before this question can be considered it is methodologically imperative to solve the problem of the relationship between time and eternity. If this relationship is denied and the meaning of being is related to eternity alone, divorced from time, then time itself (and everything temporal) is considered insignificant and empty, cut off from eternity, which is conceived as unchanging, endless continuity. What then is the use of efforts to postpone death and prolong human life, which is one of the most widely accepted aims of social progress?

But this question obviously cannot be reduced to purely quantitative considerations. A human lifetime is increased many times over if man lives each moment fruitfully, to the full. Epicurus said that a wise man does not choose the largest helping but the most attractive. The question is not only to live as long as possible but as intensively as possible, to fill each moment of life to overflowing. From a social standpoint the problem of man and time amounts to that of the rational and most intensive possible use of time. Indeed the question of the management and use of time is of exceptional importance in social and economic life. It is also a profoundly moral problem, bound up with one's view of the very meaning of human life.

It should not be thought that temporality deprives life of meaning. Man's mortality is tragic, but real tragedy is in no way pessimistic: because of the intrinsically social nature of his personality man continues to live in the personalities of others.

The truth is that man in his 'mortal' existence is linked with the real eternity of the objective world, this link being forged in the process of his creative activity. Man's involvement in labour enables him to extend his 'I' beyond the limits of his own fleeting existence. He dies but his deeds remain, his qualities continue to exist in others—the torch is handed down from one generation to the next. This is the proper formulation and the real solution of the problem of values, in which the relationship between the temporal and the eternal plays such an important role.

Technical progress and the increasing speed of transport and communications are changing people's ideas about time. Rational use not only of time at work but also of leisure time is becoming more and more important. Appreciation of the value of time is one of the main signs of a rationally organized life. A considered approach as to how to spend the available time, indeed the whole of life, is evidence of a high sense of values in a social community or in an individual.

To live in time is above all to exploit one's present to the full. But this is not to be confused with the 'eat, drink and be merry!' approach to life. Our temporal present first attains its full significance when it embodies both the past—purveyed by knowledge of the cultural heritage and possession of the material goods and working tools handed down by our forebears—and the future, being oriented towards the rising generations. It is important to have a deep and intuitive understanding of the mutual links between past, present and future and to see that the present is linked with the other two aspects of time, that each moment contains within itself the quintessence of the past and the seeds of the future.

It has been a characteristic trait of man, which has developed in the course of civilization, to strive to widen his awareness of time and to go beyond the narrow confines of the immediately experienced present. This tendency is expressed in science with its intense concern with the past, its historic sense, its ever-increasing interest in forecasting the future and its striving to discover stable laws underlying the multiplicity of events. The same tendency is also vividly expressed in art, which records contemporary reality in the novel, in sculpture and on canvas, thus capturing the intransient in the transient. For if we penetrate deeply enough into the temporal we can in it attain the eternal.

The dialectical interplay between the categories of time and eternity is fully realized in the motion of the real world. In reality we do not find a stagnant, motionless eternity over against a transient 'vanity of vanities' but a single process of constantly changing and constantly present being, that of the motion of matter, by means of which the underlying eternity of the world is realized. And there is a place in this process for man—man the inventor, man the creator.

NOTES

1. V. I. Lenin, *Collected Works*, English translation, Vol. 14, p. 175, Moscow, 1968.
2. F. A. Brown, 'The Rhythmic Nature of Animals and Plants', *American Scientist*, Vol. 47, No. 2, 1959, p. 147.
3. F. Spisani, *Significato e Struttura del Tempo*, p. 15, Bologna, Azzoguidi, 1972.
4. ibid., p. 33.
5. J. Maritain, *De Bergson à Thomas d'Aquin*, p. 30, Paris, 1947.
6. M. Heidegger, *Sein und Zeit*, Vol. I, p. 329, Halle, 1931.

7. J. J. Smart, 'The Temporal Asymmetry of the World', *Analysis*, Vol. 14, No. 4, 1954, p. 81; 'Is Time Travel Possible?', *The Journal of Philosophy* (Lancaster), Vol. 60, No. 9, 1963.
8. M. Black, 'The "Direction" of Time', *Analysis*, Vol. 19, No. 3, 1959.
9. I. Kant, *Collected Works*, Vol. 3, p. 149, Moscow, Mysl', 1964.
10. H. Bergson, *Durée et Simultanéité*, 7th ed., p. 42, Paris, Presses Universitaires de France, 1968.
11. J. Pucelle, *Le Temps*, p. 51, Paris, Presses Universitaires de France.
12. A. P. Bystrov, *Prošloe, Nastojaščee, Buduščee Čeloveka* [Man's Past, Present and Future], p. 219, Leningrad, Medgiz, 1957.
13. O. Jespersen, *Filosofija Grammatiki* [A Philosophy of Grammar], p. 302, Moscow, Foreign Literature Publishing House, 1958.
14. D. Nys, *La Notion de Temps*, p. 56, Louvain and Paris, 1925.
15. M. S. Rogovin, *Filosofskie Problemy Teorii Pamjati* [Philosophical Problems of the Theory of Memory], p. 53–68, Moscow, Vysšaja Škola, 1966.
16. P. K. Anokhin, 'Filosofskie Aspekty Teorii Funkcional'noj Sistemy [Philosophical Aspects of the Theory of the Functional System]', *Filosofskie Problemy Biologii* [Philosophical Problems in Biology], Moscow, Nauka, 1973.
17. N. A. Bernstein, *Očerki po Fiziologii Dviženij i Fiziologii Aktivnosti* [Essays on the Physiology of Motion and the Physiology of Activity], Moscow, Medicina, 1966.
18. K. Marx and F. Engels, *Collected Works* (Russian ed.), Vol. 12, p. 731, Moscow, 1958.
19. I. T. Bzhalava, *Psihologija Ustanovki i Kibernetika* [The Psychology of Purpose and Cybernetics], Moscow, Nauka, 1966.
20. For a more detailed account, see: Y. F. Askin, *Problema Vremeni. Ee Filosofskoe Istolkovanie* [The problem of Time, a Philosophical Interpretation], Moscow, Mysl', 1966. In translation in Spanish (*El Problema del Tiempo. Su Interpretación Filosófica*, Montevideo, 1968) and Portuguese (*O Problema do Tempo. Sua Interpretação Filosófica*, Rio de Janeiro, 1969).
21. F. Alquié, *Le Désir d'Éternité*, p. 103, Paris, 1947.
22. J. L. Stocks, *Time, Cause and Eternity*, p. 2, London, 1938.
23. Lenin, op. cit., English translation, Vol. 38, p. 70, Moscow, 1963.
24. J. E. McTaggart, *Philosophical Studies*, p. 155, London, 1934.
25. F. Aliquié, op. cit., p. 13.
26. Quoted by R. B. Winn in *A Concise Dictionary of Existentialism*, p. 103, New York, 1960.

TEMPORAL RELATIONS AND TEMPORAL PROPERTIES

Ted Honderich

Let us have in mind three visits to the earth of a recurrent comet, say Halley's, and one other event, the falling of a leaf. In speaking of a visit of the comet, we understand what we need not define closely, a movement of the comet in which it is near to the earth. We have, let it be supposed, some way of distinguishing each of the three visits which does not have to do with its time, and hence not with its relations in time to the other two visits. Let us suppose that we identify it by perhaps its particular visibility from England—perhaps the particular degree to which it was obscured by clouds. Thereafter we refer to one visit so identified as the *first mentioned* visit or, with exactly the same meaning, the *first* visit, and refer to the other two in the related ways.

We have certain beliefs about the four events. The first-mentioned visit of the comet, we believe, is *before* the second, the second is *simultaneous with* the falling of the leaf, and the third is *after* the second. Each of the events, then, has what may be called a temporal relation to another event. This is to say, only, that it is before, simultaneous with or after another event.[1] Taking into account only the four events and no others, each has but three temporal relations, one with each of the other events.

It would be more natural to make use of ordinary tensed verbs and thus to say, for example, that the first-mentioned visit *was* before the second, or *is happening* before it, or *will be* happening before it. However, what is in question here is a relation which seems to be independent of tenses of past, present and future. Despite the fact that the relation can be asserted in tensed sentences, and ordinarily is, they also carry further meaning. Hence, it will be best to stick to a tenseless usage.

There is also something else which we believe of our four events, something which does have entirely to do with tenses. Each event *has happened*, *is happening*, or *will happen*. Each one is *past*, *present* or *future*. Each, then, has what may be called a temporal property. This is only to say, in one way or another, that it is past, present or future.

Temporal relations are unchanging. Given that the first-mentioned visit of the comet is before the second, then it always stands in just that relation of precedence to the second. An event's temporal properties are otherwise in that they do change. The falling of the leaf, if we suppose that it is happening now, was once going to happen and soon will have happened. The event was future but ceased to be. Is present, and will be past.

Here and hereafter, it will be important to keep in mind that when temporal properties or temporal relations are mentioned, terms of convenience are being employed. They are merely invented abbreviations for more ordinary usage. To say an event has a temporal property is merely to say it is past, present or future. Similarly, to say that temporal properties of events change means only whatever we ordinarily mean when we say, in one of several possible ways, that events are in turn future, present and past. It is not being suggested that events possess properties in one of the several different manners in which items of one kind and another possess properties. It is not being suggested that temporal properties are like, for example, colour or location. So with the term 'temporal relations'. It is just an abbreviation of ordinary talk.

The fact that events have both temporal relations and also temporal properties enters into one of the fundamental disputes about time in Western philosophy. Given the nature of this book, my primary intention is to report on this controversy. My exposition of the competing doctrines, for several reasons, will be cursory and partial.[2] To advance the dispute, rather than report on it, would require a kind of close argument which here is out of place. My lesser purpose is to offer some informal argument.[3]

The problem which we shall consider involves what is meant by saying that an event is earlier than another, or stands in some other temporal relation, and what is meant by saying that an event is past, or has some other temporal property. The questions are compelling in themselves. Moreover, the answers to them are highly relevant to other mysteries about time.

A cautious analysis of temporal relations

With respect to the two temporal relations of antecedence and subsequence, as distinct from the relation of simultaneity, a part of our understanding is beyond doubt: (a) if the first-mentioned visit of the comet was before the second, then the second was not before the first; (b) if the first was before the second and the second before the third, then the first was before the third; (c) finally, it would be absurd to say that the first visit was before the first visit. There are three related truths about our understanding of the statement that one event was after rather than before another.

The two relations in question, then, are understood by us to have certain formal properties: those of asymmetry, transitivity and irreflexivity. In these three properties they are unlike certain other relations. The relation of equality, for example, is symmetrical and reflexive, and the relation of fatherhood is intransitive.

Our understanding of the temporal relations of antecedence and subsequence, however, certainly has more in it than that. This follows from the fact alone that other relations, having nothing to do with time, do have the same formal properties. We may say of items in space, perhaps the furniture in a drawing-room, that from a certain point of view one piece is to the left of another. Here too we have a relation which is asymmetrical, transitive and irreflexive.

Given the gravity with which some philosophers produce these formal properties, it needs making plain how *very* little we know about temporal relations in knowing that they have the properties. Take asymmetry. What we know about the statement that the falling of the leaf is before the third visit of the comet, given a grasp only of the asymmetry, is only that the falling is so related to the visit that the visit cannot be related in that way, whatever it may be, to the falling. We know virtually nothing of the nature of the relation. *All* we know is *explicitly* given in the statement that since the visit stands in relation R to the falling, the falling does not stand in relation R to the visit. We do not know how to understand R. Given that we have a grasp only of asymmetry, indeed, we cannot even distinguish antecedence from subsequence. We do not understand the difference between saying that the falling is before the visit rather than after it.

A second part of our understanding of the two temporal relations, it may be said, is simply that in talking of items being before or after one another, we are talking of relations which are taken to connect items of a certain kind: events. These relations are not understood to connect such items as locations or numbers. How far does this take us? We may understand an event to consist, in part, in a thing's having a certain feature. Do we, in seeing that the relation of antecedence has events as its terms, see more about the relation? The answer to that question, despite the fact that a full definition of an event may include a reference to time, seems to be yes. We could not be allowed to understand the relation of fatherhood at all, for example, if we did not understand that its terms could not be locations. Exactly what it is that we understand, however, when we see that temporal relations hold between events, is unclear.

Thirdly, talk of temporal relations may be said to have to do with change. That is, to know that our four events stand in their various temporal relations is to know of different states of the world. One state includes the first visit but none of the other three events. Another includes the second visit and the leaf's falling, but not the first or the third visit. Another state includes the third visit, but none of the other three events.

To say the first visit is before the second, then, is to talk of two different states of the world and hence of what can be called change.

This is likely to seem to contribute more to the understanding of temporal relations. However, to mention but one point of relevance, it is essential that the particular notion of change be made clear. What we are understanding by 'change' is no more than a matter of different states. In the sense in which change is involved in temporal relations, according to the suggestion, there is precisely the same kind of change in the fixed line of hills we see on the horizon. In one part the line is smooth and in another it is broken. The line has 'successive' segments or 'states'. We are not entertaining the idea of change which seems to have to do, as we may obscurely say, with a thing's coming into existence.

Is there a fourth distinct thing involved in talk of the two temporal relations? Is there something over and above their formal properties, the fact that they hold between events, and the fact that they involve a kind of change? It seems that there must be, principally for the reason that we still seem to have no distinction between antecedence and subsequence. What has been suggested is that these relations, and hence, a series of events defined in terms of these relations have a *direction*. Our own series, it may be said, has direction from the first-mentioned visit of the comet to the third. Here we come upon one of many terms of a metaphorical kind, which are to some philosophers the very guide to reflection about time. To other philosophers they are its very curse.

Those who speak of the direction of the temporal relations, and of the resulting series of events, are presumably not concerned with the fact that we commonly think of precedent events before we think of subsequent ones. We can in our thinking go the other way, and we often do. Rather, they are concerned to assert that the relations and the series *themselves* have a direction. In this, such relations and series are unlike the spatial relation of one thing's being to the left of another, or of a series of positions in space defined on the basis of that relation.

Can it be that this intuition of direction, despite its evocative nature, actually has to do only with something relatively unmysterious? Some physical processes, notably increase in entropy, are regarded as irreversible. That is, they are processes such that an initial state of a given kind is not followed by another state of the same kind. Hence, if one passes in thought one way along a series of events, one comes upon certain sequences or orderings of events which do not occur if one passes along the series the other way. What we have here, it appears, is a further characterization of temporal relations got from a feature of the series of terms defined by these relations. Much must go unsaid, but it may seem that this irreversibility will do the job of distinguishing antecedence and subsequence in the way we do.

Can it be that a search for a grander thing, in so far as the direction of temporal relations is concerned, is no more than the result of allowing one's mind to wander from the business at hand? Certainly, there seems to be a pretty impressive 'direction' in events when they are considered as future, present and past. The same is true when we simply think of events, in an ordinary and unanalytical way, as in time. However, we are not now concerned either with the temporal properties of events or with general talk that has to do with time. We are concerned only with events taken as before or after one another.

I have, as some will be aware, been guided in most of these remarks by an awareness of one tradition of philosophical thought about temporal relations. For want of a better name, and for reasons to which we shall come, it may be called the tradition of caution. We shall return to the question of temporal relations and to another tradition of thought about them. For the moment, we have the hypothesis that our statements about the precedence and subsequence of items have as their content (a) an awareness that the terms of the relations are events, (b) certain beliefs about formal properties, (c) a belief having to do with what is called change, and (d) some belief or other about direction.

This sketch of one possible analysis of statements about temporal relations is lamentably unfinished. It raises many questions which must go unanswered. However, it is worth noticing that it has a further feature so far unmentioned. What I have in mind is (e) that this analysis of statements about temporal *relations* does not have recourse to temporal *properties*. It does not attempt to explain the belief that one visit of the comet is before or after another by bringing in familiar notions of past, present or future. The same is true, incidentally, of a closely related analysis of the temporal relation of simultaneity, which we have left untouched in all of this.

A cautious analysis of temporal properties

Let us turn now to our beliefs about the temporal properties of our four events. Let us, for the purpose of reflection, suppose that the second visit of the comet is now happening, the first is past, and the third is to come. If we believe these things, what is it that we believe? If we are guided by the philosophical tradition of caution already mentioned, we shall answer that the meaning of such statements about temporal *properties* has to do with temporal *relations* conceived in the way that we have just conceived them.

The answer, in one clear version, can be given very quickly. It is that my statement, which we may call S_1, that the second-mentioned visit is now happening, amounts to this: the second-mentioned visit is simultaneous with my utterance of the statement S_1. We shall say that the statement, S_2, that the first visit is now past, amounts to this: the first visit is before my utterance of the statement S_2. As for the statement S_3, that the third visit is to come, or is in the future, it amounts to a claim about the third visit's being after my utterance of S_3. All this, at least after a moment's reflection, may seem natural and persuasive.

If we bring together the analyses of temporal relations and temporal properties, we have a view of our talk of time. Let us notice two principal features of the view. Past, present and future are reduced to the relations of precedence, simultaneity and subsequence. We have, in all of our talk about time, no more than ideas of the kind supplied with respect to temporal relations. All of our talk about time, including talk of temporal properties, is reduced to the particular ideas which we supposed ourselves to have in speaking of events as before other events, simultaneous with them, or after them. The relations which we are said to have in mind in speaking of past, present and future are different from others of their kind only in one way: they have as one term a linguistic event, and thus an event dependent upon the existence of consciousness.

There is then a second principal feature of the view which we are entertaining. Given the particular understanding of temporal properties, there would be no such properties in a world without consciousness. The point is not, of course, that the properties would exist, but that there would be no awareness of them or beliefs about them. There would in fact be no such properties, since the existence of such properties depends on the existence of consciousness.

More explicitly, the view we are entertaining has the consequence that to say an event is now happening is to assert or presuppose the existence of consciousness. Given that view, what we believe in believing that the second-mentioned visit of the comet is happening now, would not be true if no one were here to make statement S_1 or some version of it. If the world were without conscious inhabitants, the second-mentioned visit of the comet would indeed be after the first, simultaneous with the falling of the leaf, and before the third visit. If what we mean by 'now' and our tensed utterances is what we are supposing we mean, it would not be true that the second visit was happening now, or, as we ordinarily say, *happening*. It would not be true that the first-mentioned visit had happened or was past, or that the third visit was to happen or was in the future.

There is no doubt that when this second feature of the analysis before us is made clear, some people will be quick to conclude that the analysis is unutterably mistaken. They may also conclude that arguments against it can easily be discovered. There is something to be said for restraint, however. Let us turn to the alternative tradition, and first to some reflection on temporal properties rather than relations.

An affirmative analysis of the temporal property of presentness

When we depart from our ordinary thoughts and utterances, in which items are taken to be past, present or future, and direct our attention to the substantive term, 'the present', we come upon a host of quite familiar usages. It is natural to say, for example, that the present passes, or passes away. We may indeed become enthusiastic and declare, with one philosopher, that the present has 'the jerky or whooshy quality of transience'. Such utterances and declamations, no doubt, are distantly related to some truth. They may also seduce one into confusion.

Some philosophers, it appears, have been led by them to conceive of the present as a *thing*, and indeed a thing distantly related to the rolling stock of British Rail. The principal difficulty with such ideas, however expressed, is that all movement in the end consists in, or at least requires, a thing's being in different places at different times. If 'the present' is a kind of moving thing, in what time does it do so? We started with the present, or, as we can say, present time, and we must now have another time in which it makes its progress. Does that further time also have a moving present?

Let us leave all that without further ado. Let us not forget, however, our familiar usages about 'the present'. We are not so well supplied with kinds of data about time that we can ignore any we have. Let us keep the familiarities at hand, forget about time as a moving thing, and return to the statement that the second visit of the comet is now occurring. What does it mean?

Perhaps the truism that the second visit is neither future nor past suggests a line of reflection. It makes some sense or other to say that the future does not yet exist, and the past no longer exists. Can we then attach sense to the statements that future events do not yet exist and that past events no longer exist? If so, we may move to the obscure speculation that to say that the second visit of the comet is now occurring is to say that it now exists.

Whatever else we have, we now appear to be in line with our familiar usages. We need not take the step of ignoring our notion that the present passes. We can attempt to understand it as having to do with such matters as a thing's coming into existence, being in it, and going out of it.

Still, we are in difficulty. If we consider one of the things just said, that past events are events which no longer exist, one of our embarrassments may be that it seems to amount to this: past events do not exist *now*. We shall not have made much progress if we explain our use of 'past' by introducing, in the explanation, a use of 'now'. We wish, somehow, to give an explanation of all the terms by which we ascribe temporal properties. We do not wish merely to make connexions between the terms, to 'define' one by means of the other, but rather to get outside the ring of terms. Similarly, we have just supposed that to say the second-mentioned visit of the comet is occurring now is to say that it, the visit, now exists. How much good is that? The difficulty is plain circularity. 'Now' turns up once too often.

Also, and differently, there must be some doubt that we have made a clear advance by moving from talk of the visit's *happening* to talk of the visit's *existing*. We need not be too perturbed by the response of the plain man, or the plain philosopher, that it is things which exist, rather than events. We can admit that the usage is unusual, and attempt to justify it. Let us look at this difficulty first and then return to the matter of circularity.

How does it come about, as it does, that philosophers are led to assertions about time like the one we are considering? The explanation often has a great deal to do with the cautious analysis of temporal properties at which we have looked already. That analysis issues in a particular understanding of the statement that the second visit of the comet is now happening. The happening of the visit is made into a matter of a simultaneity relation. This, it is felt, is wonderfully inadequate. It *is* true that when we say truly that the second visit is now happening, our utterance is simultaneous with the visit. Surely, however, that is not what we are saying, or all of it. Something other than that is being said.

In order to attempt to express what seems to be meant by the statement that the visit is happening now, the philosophers in question have recourse, as we have had, to the declaration that it now exists. We need not look at several variations on this theme.

The trouble is that in so far as analysis is concerned, we get very little from this talk of existence. Until more is said, the clear content that it has seems to amount to two implications. It is implied that to say the second visit is happening now is not to say something, or not only to say something, about a simultaneity relation. It is implied that the present happening of events would continue if we were not here to be aware of events.

What *are* we to understand that the statement about the second visit *states*? The only answer I can see, one to which we might as well come sooner rather than later, is that in saying of the second visit that it is now happening, we are saying something of the visit which we all understand but which is not open to further analysis. We here have an unanalysable or primitive notion, expressed by such descriptions of something as 'happening now'.

If we take up this view, we shall certainly have to put up with the standard retort made to anyone who asserts about a controversial notion that it is unanalysable or primitive. The retort is that saying the notion is primitive is no more than an admission of failure in the task of analysis. It is to be replied, perhaps, that some such 'failures' consist in the perception of truth. The existence of certain primitive notions in our conceptual scheme is undeniable and perhaps has never been denied. Is there adequate reason for denying that we have another one here? Certainly time is in some sense basic to our perception and reflection. What more likely place to find an unanalysable concept?

What we have, then, is the hypothesis that to say the second visit is happening now, or simply happening, is to say something about the visit which is understood by all and which is in a certain sense simple or fundamental and thus not open to reduction. And, since we have not got an analysis of statements about present events, we have not got a circular analysis of them. *That* objection is no longer a possibility. The circularity that was apparent in our attempted analysis, however, is not without importance. It, like the many circularities which emerge in discussion of time, is an indication that we *are* dealing with a primitive concept. As for circularity that may arise in analyses of statements about past and future, we shall return to the matter.

One of two questions that remain about our hypothesis as to statements about present events has to do with the formulation of the hypothesis. Shall we persist in saying that the hypothesis is that such assertions as the one about the second visit ascribe existence to events? Perhaps there is no harm in this so long as we know what we are up to. The hypothesis, if we choose to *assert* the implications mentioned above, consists in three propositions: (a) statements about temporal properties do not have to do with, or only to do with, simultaneity relations; (b) such statements would be true if there were no consciousness; (c) such statements rest on a primitive notion of existence.

The other question is this one: are we in fact engaged in an absurd enterprise of inventing a notion that is simply redundant? 'We all agree', someone may say, 'that the second visit of the comet is now happening. That is, the comet *is near the earth.* The comet has that feature. Can we really suppose sensibly that on top of that, something else is importantly true, something which you refer to by saying that the visit has "existence"?' The proper answer must be that we are not *adding* anything. What we are doing is supposing that to say the second visit is now happening, or that it is happening, or that the comet is (now) near the earth, or that the comet (now) has a feature, *is* to make use of a notion for which there is no analysis.

We have another account, then, of one class of statements about temporal properties, those that have to do with present events. It is an account in accordance with the second tradition of philosophical thought about time. The first tradition, that of caution, has the distinction among others of being informed by a reluctance to take up anything about which clear things cannot be said. The second tradition expresses, among other things, an attitude of affirmation. We may call it the tradition of affirmation. What is affirmed is principally one conviction about the present, and it is affirmed in spite of the fact that nothing of an analytical kind can be said about it.

An affirmative analysis of past and future and of temporal relations

To continue with temporal properties, we must have some view of past and future. It seems sometimes to be supposed by philosophers of the second tradition that all of our talk of temporal properties can be explained, if that is the right word, by seeing that it rests on a notion of 'becoming', or 'passage' or 'real transiency'. Here we appear to have bundled up into one bag all of past, present and future. One may wonder, of course, if these terms are intended to have to do only with talk of events being *present*. That is, one may wonder if they are merely other ways of referring to the primitive notion with which we have been concerned. If so, their intended use is entirely analogous to our use of the term 'existence'. The suspicion that more is intended, that the terms in question are thought to enlighten us about all of past, present and future, is sometimes reinforced by the fact that nothing at all is said explicitly about past and future. No separate account whatever is given of past and future.

Be that as it may, and to come to something clearer, it would be eccentric to claim that we have a single primitive notion of past, present and future. After all, we do distinguish these three things. It would also be unacceptable, surely, to call into being two more primitive notions, one for past and one for future. In so far as there is an argument for the assertion that a notion is primitive, it must consist partly in the fact that no tolerable analysis for it can be found. Given that we have what we seem to have, including a notion of events as present, this turns out to be untrue of past and future.

One possibility about past and future that must come to mind is that they be analysed in terms of the notion of the present and the notions of precedence and subsequence. That is, we make use of those two temporal relations, along with our concept of the present, in order to explain past and future. What we consider, obviously, is that to say an event is past is to say that it is before present events, and to say an event is future is to say that it is after present events.

Let us reserve judgement on that possibility and turn our attention directly to temporal relations. Let us reconsider their analysis. Philosophers in the tradition of affirmation have usually been unsatisfied with the kind of analysis given of temporal relations in the tradition of caution. They take it to be mistaken, or anyway inadequate, to describe the relations of antecedence and subsequence by way of the nature of their terms, formal properties, and, in certain limited senses, change and direction.

Their move at this point is also predictable enough. It is said that part or indeed all of what we mean in saying the second-mentioned visit of the comet is after the first, is that when the first visit is present the second is future, and that when the second is present the first is past. A similar line is followed with the relation of antecedence and that of simultaneity.

Whatever else is to be said of this analysis of temporal relations, it is clear that it has one consequence. If we adopt this analysis of temporal relations, we must give up the possibility noticed a moment ago with respect to the temporal properties of past and future. If we adopt both things, we shall again have a circularity. We shall be offering an analysis of past and future which depends in part on the notions of antecedence and subsequence, but those latter notions will themselves be explained partly in terms of past and future.

There seems to be a way out of this difficulty, although one cannot follow it with great confidence.

If, like the philosophers of the tradition of affirmation, we are inclined to think the cautious analysis of temporal relations is mistaken or inadequate, what is the root of our inclination? It is, in a sentence, that what is said of temporal relations makes them too much akin to spatial relations. It would be mistaken, for obvious reasons, to claim that the cautious analysis of temporal relations does not make them at all different from spatial relations. None the less, one feels there is not enough difference.

To put the question one way, what can be added in order to get *time* into the analysis of temporal relations? Let us first recall that items in temporal relations form a series. This fact, which is obvious enough, can be explained partly in terms of the formal properties of the relations. The four events with which we have been concerned are members of this series, which also includes events before the first-mentioned visit of the comet and after the third-mentioned visit. We are inclined to regard the series as in a way infinite. Be that as it may, and to come to the essential point, let us say of this series that some one event in it, and all events simultaneous with that event, are *occurring* or are *present*. That is, they are such that the predicates which express our primitive notion are true of them. In short, then, we add to our account of temporal relations that they hold between events which form a series of which some member-events are present.

This additional characterization of temporal relations is to be distinguished from something we have considered already: the characterization of them as holding between *events*. At any rate, it can and is to be distinguished from anything that might be meant by the latter characterization within the cautious tradition. There, presentness is restricted to a matter of simultaneity relations. Whatever may there be added to the analysis of temporal relations by specifying that they hold between events, it cannot be our current idea. All that talk of events being present is allowed to mean in the cautious tradition is something about simultaneity relations. Nothing would be gained by saying, in explanation of the relations of antecedence, simultaneity and subsequence, that they hold between terms which can enter into certain simultaneity relations.

One advantage of the suggestion that temporal relations be understood partly as relations between items which form a series having a certain feature is the advantage that we can now complete our analysis of temporal *qualities* without circularity. We are enabled to say that past events are to be understood as events before present events, without the embarrassment that the notion of antecedence has been explained partly in terms of past events. Similarly, we seem to be enabled to say that future events are to be understood as events after present events, without the embarrassment that the notion of subsequence has been explained partly in terms of future events.

Summary

Let us sum up the accounts of temporal relations and temporal properties. The account of temporal *relations*, taken from the tradition of caution, is to the effect that such a statement as this one, that the first-mentioned visit of the comet is before the second, conveys four things. The related terms are events, the relation has certain formal properties, implies the existence of a kind of change, and has what is called direction. The account of temporal *properties* taken from the tradition of caution is to the effect that such a statement as this one, that the second visit of the comet is now occurring, conveys that the visit is simultaneous with the utterance of the statement.

Temporal *relations*, given the account suggested a moment ago, one within the tradition of affirmation, are to be understood in terms of the four features plus the further one that the terms of the relations form a series such that some of its member-events are present. As for temporal *properties*, the statement that the second visit is now happening is taken to involve the use of an unanalysable notion. The statement that the third visit is to come is taken to mean that it is after present events, and the statement that the first visit has occurred already is taken to mean that it is before present events.

Speculations

To repeat something said in the beginning, I do not suppose that a decision can be made between these doctrines on the basis of what has been said here. We have before us only hurried propositions and inconclusive reflections. None the less, if a verdict cannot be reached, we may at least speculate about one.

The cautious account of temporal properties seems unpersuasive. Indeed, it seems open to something like a disproof. Take one example, the cautious analysis of the statement S_3, that the third visit has not yet happened, or is future. The analysis must surely be mistaken. If I make the statement, what it means according to the analysis is that the third visit is after my utterance. The statement means, in short, that two events are in the temporal relation of subsequence. Temporal relations, as we know, do not change. Hence, when the third visit *is happening*, and when it is *past*, it will remain true of it that it is after my utterance of statement S_3. But surely it cannot be that what is meant by S_3 is something that will be true when the third visit is present or past.

We may notice at this point, if only in passing, a related doctrine which is not open to this objection, whatever else may be said of it. What it amounts to, in terms of our example, is that S_3 conveys that the third visit of the comet is after a *present mental event* of mine, one associated with my utterance. Similarly, the statement S_1, that the second visit is now happening, conveys that the second visit is simultaneous with a *present mental event* of mine, one associated with my utterance of S_1. Such mental events, it is allowed, are *present* in a sense in which physical events are not present. What can be meant by this? It is fair to observe, I think, that no answer to this question is given. One is driven to suppose, in the absence of any instruction to the contrary, that when mental events are allowed to be present, what is being relied upon is a primitive notion.

There are several reasons why this particular view does not command attention. To mention but one, the admission that mental events have a temporal property, conceived in a certain way, seems to lead inevitably to the admission that physical events also have this property. This is so, essentially, because of the connexion between mind and body. The proponents of the view have struggled manfully to avoid this further admission. In my opinion, they have not succeeded. Hence the view, although encumbered with propositions about simultaneity, comes to rest in the opposing tradition of affirmation.

Putting aside temporal properties, the tradition of caution also seems open to question in its analysis of temporal relations. The objection here must be less clear-cut. Philosophical inquiries of the kind in which we have been engaged are directed, importantly, by pre-philosophical convictions, perceptions, guesses and so on. Analyses, in the end, are assessed against these things. No one will suppose, on reflection, that all such convictions and what-not are sacrosanct. Some of them can be rejected for the reason that they do conflict with the outcomes of philosophical inquiry. Still, to put the point far too quickly, it is difficult to escape the feeling that the cautious analysis of temporal relations is the loser in its real enough conflict with pre-philosophical commitments. One persists in the view that we do mean a good deal more in our temporal-relation utterances than is allowed by the analysis.

To turn to the tradition of affirmation, and more particularly to the particular affirmative doctrine at which we ourselves arrived, it appears that it is not open to *refutation*, at least in any economical way. Also, it appears not to be involved in a losing conflict with pre-philosophical commitments. That is all very well, but it is not all that matters. The foundation of the doctrine, obviously, is the claim that there exists a certain notion that is beyond analysis. After the claim was first introduced, we did notice the possible reply that such claims about primitive notions are merely confessions of failure. There is, as we saw, the possibility of the brave rejoinder that the 'failure' may be a happy one since 'success', or the production of an analysis, would simply be a mistake. However, a brave rejoinder is not to be confused with coercive argument. There remains the possibility that an *analysis*, something deserving of the name, can be given of statements about temporal properties.

NOTES

1. Throughout the essay, I speak of the three principal temporal relations only. However, there are obviously more than three such relations. An event is also *just before* another, *long before* another, *two minutes after* another, *a century after* another, and so on. I ignore these further relations, which raise no special problems for me.
2. More can be learned of these doctrines from *The Philosophy of Time, a Collection of Essays* (London, 1968), edited by Richard M. Gale. The book contains essays by different philosophers and also an extensive bibliography.
3. I am grateful to my colleagues, Dr Malcolm Budd and Dr John Watkins, for comments.

TIME AND ITS SECRET IN LATIN AMERICA

Saul Karsz

This study attempts to analyse psychological and moral attitudes to time in Latin America—much too vast a subject to cover fully, but certain of the outlines can be traced, and illustrated by significant names and trends. Detailed coverage would inevitably involve serious oversights: we prefer to single out, and suggest a schema which might lead on to research in greater depth.

I

Are Hispano-America and Latin America synonyms? As distinguished from English-speaking North America, Hispano-America means the vast sub-continent to which the Spanish—and in the case of Brazil, the Portuguese—conquest gave a unity which included great variations from one part to another. Hispano-America as such began to disappear with the wars of independence and the military, economic and cultural influence of England, the United States and France. Before the end of the nineteenth century, the name was no longer correct, and could even be misleading in regard to the factors and forces which helped to shape the continent.

Latin America proper has existed for a century and a half without, however, altogether superseding Hispano-America. There was in fact a fierce struggle between the two throughout the first half of the nineteenth century, each having partisans in both America and Europe.

Five major social groups were distinguishable: Indians, Spanish (or Portuguese), black slaves, half-castes and creoles, each with a particular attitude to time in keeping with its social and institutional role, but all sharing a common denominator—the Catholicism imported with the conquest.

Between Catholicism and the Protestant and Calvinist faiths of North

America the difference is fundamental. In the eighteenth and nineteenth centuries, the latter demanded tangible proof of a relationship with the God: no one is saved once and for all but by individual effort, by work and, above all, by social and financial success, must prove himself worthy of the world that God has created. Hence, time means the austere self-discipline of work, untiring accumulation, and material efficiency, the premise and sign of the heavenly destiny of the elect. This is pragmatic time, aggressive towards nature and in practical day-to-day relationships, charitable towards equals within the community. It was the time of those who had everything to forget, i.e. the European past, and everything to gain, i.e. the American present and future.

The Catholic attitude in Latin America was quite different. Time was perceived as against an all-encompassing and unbroken eternity which historical time imitates but can copy only fleetingly and erratically. Time was thus waiting transition, for its beginning and its end were elsewhere: morally in eternity, politically and economically in the mother country. Latin American time referred to something outside itself; not itself decisive, but rather illustrative of a beyond located both in the past and in the future, but from which the present was missing. This colonial time, from the seventeenth century to the nineteenth, was characteristic of Hispano-America.

Colonial time implies disorientation. For the indigenous populations, organized loosely or (as the Aztecs and Incas) in empires, the arrival of the Europeans meant destruction, brutal press-ganging into forced labour and partial evangelization. Their traditions, broken by the *conquistadores*, survived beneath their conversion to Christianity. Cut off by the conquest from their past, the present existed for South American Indians only as one of the metamorphoses of a time which they conceived as cyclic and self-repeating. The colonial present was unreal, a dream-time awaiting the impossible return of their former life. Desperate uprisings, quickly crushed, failed to mend the broken circle. Time broke into impermeable blocks, impregnated with pessimism and resignation.

For the *conquistadores*, colonial time was a necessary and rewarding interlude before the return to the mother country, to serve which they had ventured forth to a continent where time appeared as immeasurable as the land itself, punctuated only by news from Europe, and the business of administering local affairs. Compared to anything they had known, everything seemed new and inordinate. To quote Martínez Estrada, in his *Radiografía de la Pampa* (1933): ‘Property, authority, human relations, trade, family life were all subject to unforeseeable changes, like plants which, newly transplanted, may take root or may die.’ America represented a time of challenge, an untamed, unconfined and disturbing dimension which had to be mastered for Spain or Portugal. For this, the *conquistador* had to concentrate on the new world and settle there, while his thoughts were elsewhere. He had indeed a rich past, but he could defend it only by coming to terms with the present: faithful to a real but increasingly remote time, he had to live in a time that was provisional, but immediate and real. Via this contradiction, the *conquistador* became Hispano-American.

The blacks brought to America as slaves by the conquerors remained isolated from local traditions, both indigenous and European. For them, time appeared to have stopped once and for all. Ruthlessly exploited, they kept alive in their festivities and songs the memory of another age, when time had flowed from the past towards an open future, whereas in their American present, time had come to a standstill on a continent about which they knew nothing. To escape, they could either nurse their nostalgia for a past which was happy because it was their own, or revolt and try to revive it in the here and now. Their festivities were thus something more than entertainment. Melancholy allegories of the past, they paved the way for those reprisals for stolen time whose saga and failure were recounted with magical realism by Alejo Carpentier in *El Siglo de las Luces* (1962).

The Creoles, born of the *conquistadores* in America, knew no time but the present and desired only the future. The Creole could claim forebears and invoke traditions (that he had never experienced), but he had no past of his own. His past lacked substance. He jealously cultivated it, not so much to establish his identity as to remain distinct, in time present, from Indians, slaves, half-castes and, in a way, from the conquerors. But this present, his reality, was hollow; far from repeating the present, the future should modify it, for the Creole wished to be fully master of American time and space. He looked forward to the future as *his* future, identified with that of the continent. Time for the Creole was thus open-ended, waiting to be reshaped to the measure of men for whom America must be reconquered and Americanized.

This description might suggest a mere haphazard juxtaposition of different relations to time. This is not the case, and there is indeed a central unifying factor. We noted that each of the social groups identified with a time situated elsewhere, each different from the other, but all had a common destiny in the colonial present. This might be a subject of regret or celebration, but this present was decisive and dramatic; it crystallized their hopes and projects, and was the real point of reference for one and all.

At the beginning of the nineteenth century a kind of coagulation of time accordingly occurred, and took on unprecedented importance: it was time which conferred *definition* and thereby undermined two complementary images: the fictitious eternity of colonial time, and the mother countries' unshakeable view of the colonial situation.

Specifically Latin American time can be understood only by reference to this new time-value and the undermining of these two images.

II

Latin American time is also *conflictual time* because of the prolonged period of wars of independence under Creole leadership against the metropolitan countries and their local partisans, and subsequently, because of foreign interventions and various social and political struggles.

Conflictual time was destructive but also creative in several ways, and it affected all aspects of private and social life by helping to merge individual aspirations for the first time in a collective ideal—a reorganization of attitudes, indeed a veritable mobilization of time. Time speeds up, becomes an issue, a theatre of conflicts which gradually permeate society, time syncopated by specifically Latin American imperatives. The whole continent awakens, like Kant, from dogmatic slumber.

The radical changes throughout Latin America introduced a new scansion of time. There was now a 'before', no longer the past of each but colonial time overturned, a 'now', in relation to which everyone must define himself, and an uneasy but exciting 'future'.

In other words, time began to be viewed in terms of history. For example, the abolition of slavery was proclaimed in the Rio de la Plata area in the following terms: 'Henceforth, the sons of slaves born in America shall be free.' Space was restructured with time as a parameter, and the appropriation of time underwrote liberation. These three terms—space, time and liberation—are three aspects of a single process which has remained an inevitable constant in all of Latin America down to our own times.

III

The Enlightenment was certainly one of the first, if not the first, intellectual movements to have specifically Latin American representatives. Thinkers such as Miranda, Monteagudo, Nariño and Moreno, who also played political roles, saw the relevance to Latin American conditions in the first decades of the nineteenth century of the ideals of the French Revolution. The latter they regarded both as a specific historical event and a flowering of eternal human principles, freed at long last from their straitjacket.

The attitude to time of the Latin American Enlightenment was grounded on the idea of natural rights and universal reason. Relying on the former, they deemed all men equal, equally entitled to dignity. Relying on the latter, they considered the past to be a tissue of errors, not to be repeated, and the future as an adventure, of which universal reason would alone be judge. The present was to consummate the break with tradition and take institutions from a colonial to a rational phase under the guidance of enlightened minds.

The latter wanted to recast the diversity of subcultures and peoples—

indigenous, black, half-caste, Creole, Spanish and Portuguese—in the single mould of universal reason; a single time-rhythm would elevate Latin America to the ideals of the French Revolution. To live according to one's time meant adopting Europe's time, rationality and life style.

Seen in this way, time is open-ended, moving towards a specific model. Far from being measured by memory or tradition, time adapts to the discovery of new rhythms and hurries on as rapidly as possible to a necessary and inevitable future. Underlying this conception, however, is a rejection both of colonial time and the theologically inspired thinking from which it stems, and of all attitudes to time—including Indian, half-caste and provincial—not sanctioned by universal reason. Hence the permanent opposition between upholders of this conception and those who hold to colonial time or defend Indian and half-caste rhythms and, more generally, the rhythms of the seasons and the earth.

IV

Other conceptions, mostly romantic and naturalist, were in vogue between 1835 and 1855 (cf. E. Echeverría (Argentina) and F. Bilbao (Chile)).

Historically, this was the period when most of the Latin American nation-States were coming into being. Although certain of its basic tenets were accepted, the Utopian character of Enlightenment time was attacked: Latin America could not, by a sudden transition, catch up with the modern European nations, but needed a laborious process of education. This latter idea provided the highroad to a new attitude, based on the ideas of order and of law.

Order guarantees the normal passage of time—normal in the sense that the time of local wars and *caudillo* incursions had to be superseded by a centralized time, set by the cities but observed equally in the villages, a national time which effaced regional rhythms. The law ensures this and effectively implements it. Education provides the highway because the *insubstantial*, anarchic, gregarious time of the uncultured periods must be superseded by *time in a deeper sense*, time still to be discovered and respected, when the individual will be fully master of himself. This mastery education must provide.

For the Romantics, time, to be authentic, had to be something inward and accepted, but not in pure contemplation only. On the contrary, they saw time as materially exteriorized, anchored in institutions and in practices, as the necessary accomplishment of inner time. Inner time had to be exteriorized to exist at all, but remained the module and yardstick of exteriorized time, its basis essentially individual and *self-centred*.

The Romantics regarded time positively for its potential power of producing a classical epic that could adorn any tradition. Seeking, behind historical time, an exteriorization of inner time, the Latin

American Romantics started a certain indigenist, folkloric fashion that celebrated not so much a particular historical episode as a happy balance between a way of life and a certain self-mastery—see, for example, M. Hernández' *Martín Fierro*, or R. Güiraldes' *Don Segundo Sombra* (1926). However, only *Martín Fierro* gets beyond the Romantic influence to depict, in the tribulations and persecutions of a *gaucho* archetype of a people, a temporality unknown to or repressed by the 'civilized'.

The Latin American Romantics were thus positive towards time only in so far as it offered a form of harmony and reconciliation this side of and beyond breaks and upheavals. They pioneered a *classicism* which has left a lasting imprint on Latin American culture: the past, honoured as myth, overhung by a future that will bring a return in glory, the present accorded the specific role of filling the breach between nostalgia for what was and longing for that which does not yet exist.

V

Positivist currents in Latin America, though frequently underestimated, were of undoubted importance. The influence of the positivists, who were contemporaries of the Romantics, became predominant in the second half of the nineteenth century, dying out only around the 1920s. In Brazil, veritable sects of positivists, more religious than philosophical, continued almost to our own times. Positivism is associated with the beginnings of industrialization, and with the greater role of science and physics and their effects on the understanding of man.

Latin American positivism, perhaps more than other currents of thought already mentioned, was no mere local adaptation of theses already known in Europe. It made its own way, as can be seen from a careful reading of the works of Saco, Lastarria, Sarmiento, Alberdi, Arosemena, Sierra and Ingenieros.

Comte's classification of the stages of man's evolution was reflected, sometimes anticipated, in currents of positivist thought in Latin America. The time had come for them, too, to abandon the metaphysical stage of the Enlightenment and Romanticism that had superseded the colonial theological stage, and to enter the positive stage. Sierra was still more explicit: the military age must give way to the industrial age. A new attitude to time had arrived that had no need of Providence, was concerned with concrete historical time and whose key concept was progress.

Providence was rejected in their constantly reasserted determination to get beyond colonial time and reliance on extra-human agencies in the ordering of human affairs—a necessary but now superseded stage in the inexorable onward march of mankind. They must follow the lead given by Europe and the United States. The rhythm and requirements of human undertakings, industrial in particular, set time as its optimum. Since the positivists admitted no Providence to fore-ordain man's success, no Providence might set the pace of human time either. The only transcendence was in the will to order the present in terms of what the future should be.

Banish the past, affirm the present, order the future: progress is the key to the positivist attitude to time. Positivists might variously stress the need to civilize, or to populate, or to educate; but all these ideas imply that time is evolutionary. They were hostile to revolutionary change, and asserted that time has an objective, immanent sense which has to be discovered and submitted to: time proceeds in a straight line, from lower to higher (cf. Sarmiento's *Facundo*). This was why the positivists considered any reversion to former stages barbaric and any attempt to skip a phase as anarchy. Time, perfectibility and normality are inseparably linked (cf. the writings of Ingenieros on delinquency, which repeatedly refer to *time's marginals*—barbarians and anarchists).

Finally, to be free meant taking the Anglo-Saxon pragmatic view of time, with its accent on efficiency that Latin American positivists particularly liked. But they were also conscious—especially in Mexico—of the budding antagonism between the United States and Latin America.

Progress, in any case exists, since time follows a strictly logical course and has a precise end: facts. To determine these facts and show that time present really corresponds with its own immanent version, Arosemena, a writer from Panama, announced in 1840 his 'science of factology', which was to herald a 'new era'.

The respect for facts which gives positivism its strength also defines its limits. Paradoxically enough, their relentless pursuit of concrete time made them defenders of abstract time. There is surely nothing more unrealistic than their preoccupation with pure facts, without presuppositions of any kind; not because they cannot be known, but because getting to know them depends on the soundness of the theoretical, and not only the operational, hypotheses involved. By using the concept of progress to identify concrete time and the empirical succession of events, the positivists reinstated a transcendent time which they claimed to get rid of. Their view of temporality was profoundly theological and theocratic.

The idealist and spiritual schools which rejected the notion of progress and dissociated time from history, intelligibility from praxis, were not slow to attack. The history of ideas in Latin America can, to a large extent, be seen as a conflict, lasting down to the present day, between positivist and Romantic concepts of temporality.

VI

Classifications which set out to be exhaustive risk overlooking something vital. Full justice is not done above to the wealth of Latin American conceptions of time; these were also expressed in other theoretical and practical forms which we may term 'militant'. They have a long tradition, and now, in this second half of the twentieth century, undoubtedly embody what is most universal and vital in Latin America. As early as 1542, Father Bartolomé de Las Casas took up the Indians' defence, and so started a tradition of polemical writing on social conditions among the masses and their real experience of time.

In the nineteenth century, Simón Bolívar, man of action and polemicist, fought for the geographical and political unification of Latin America, knowing that on unity its future depended; the present was when Latin America would decide its destiny for a long time to come. The failure of his plans left the way open for internecine struggles, and reinforced local differences.

It is in the life and works of the Cuban José Martí that the militant conception of time finds its best expression. Martí believed that a comprehensive knowledge of the past was necessary to effective intervention in the present. Such knowledge should be thorough but not bookish. In practical terms, getting back to origins meant recuperating Indian, black and half-caste time, which had survived secretly and apart from the officially vaunted traditions that had no basis, moral or material, without them. For Martí, the future would be a freedom struggle of those who had been dispossessed of their past and cheated of the present—no mere local affair, but involving the whole continent and all men.

These three examples will suffice for a brief illustration of the militant conception: time grasped on the basis of questioning Latin America, its forces and conditions, as they stand, so as to build a future that will be, not imposed, but chosen.

VII

Psychological and moral views on time in Latin America are not confined to speculative and philosophical texts. The Catholic Church is also actively concerned, 'traditional' and 'dissident' parties debating against the background of a common theology. Part of the clergy and of the laity are seeking to redefine the concept of time, and by getting away from colonial ideas, to bring the Gospel message into everyday contemporary living.

They affirm that this message is modern and are determined that its universality shall not impede its practical present fulfilment. Knowing that time is deeply conflictual in Latin America, they want to enmesh the religious ideals of the Scriptures with present-day conditions. Eternity, supreme factor in religious questions, is reinserted in the here and now, not so that it may be absorbed into the present but to transfigure the present and its pressing cares. Everyday reality is not at odds with the eternal, but is a branch and manifestation of it.

The poet and priest E. Cardenal (born in Nicaragua in 1925) offers a good example of this attitude towards time. He is one of the founders of the 'exteriorist' school of poetry which tries to relate the world and its conflicts to internal, subjective rhythms, and heal the rift, religious in origin, between internal and external time, between time as individual hope and time as collective fulfilment. This is the evangelization of time, which becomes moral, ceases to be profane, secondary and inconsistent, and is recovered as time transcendent. In this form, the dissident conception of time does not endanger the deep theological unity of Latin American Catholicism.

The various forms of violence in Latin America today are not unconnected with these new attitudes to time, inside and outside the Church. Independently of its nature, widespread violence imposes a radically new attitude to time, since the concepts of 'death' and 'normal times' are drastically altered. Death is no longer the biological and statistical termination of the existence of an individual who has been child, youth and adult, but an immediate possibility of a brutal end to any vital project. Time is thus out of joint, and its normal course, measured and harmonious, is at any moment liable to be disrupted by the unexpected. Death, seemingly become commonplace, gives time the dramatic stature of adventure and challenge.

It may be sad that it should be precisely death that restores this stature, but it is none the less true that the generalization of violence in various forms, resulting from conflicts on fundamental issues in Latin America is impregnating all aspects of social and private life. Better than any analysis, this daily impregnation proves that in Latin America time can be lived only in terms of history. Psychological and moral attitudes to time cannot be dissociated theoretically or in practice from this crucial link with history.

VIII

Novelists, poets and film-makers have also concerned themselves with time and attitudes to it. In frescos profoundly influenced by Mexican history, painters such as Rivera and Siqueiros have personified specifically Latin American time in great, noble figures, massive and tender, for whom time is a colossal undertaking with *building* as its keynote.

Theatre experiments, mainly by non-professional but ephemeral groups, stage legends and historical episodes that invite active involvement in something which is not just acted before the spectator but a happening in which he is himself included. The play's time and living time coalesce, each reinforcing and illuminating the other.

This attitude to time, this '*presentifying*' *of the past*, informs much of current Latin American literature. Myths and episodes are evoked not so much in nostalgia for some idyllic time as to bring to the surface buried elements which continue to be decisive now. The concern is with the present, considered both as an end-product and the occasion of a possible demystification.

This, however, does not represent the triumph of crude naturalism or materialism; the idea is that the past, far from being taken as a model, constitutes one of the raw materials which the present must transform. A whole literature is devoted to this revaluing, or rather, recognition of suppressed times, particularly Indian and peasant times. *Metal del Diablo* (1946) by A. Céspedes, and *El Precio del Estaño* (1960) by T. Terán, both Bolivians, showing miners' time, darkness fitfully punctuated by explosions a hundred metres underground, are examples. The Peruvian authors Matto de Turner (in *Pájaros sin Nido*, 1889) and C. Vallejo (in *Tungsteno*, 1931), and the Paraguayan writer R. Roa Bastos (in *Hijo de Hombre*, 1959) also deal with what we have called 'militant' time.

The Latin American attitude to time, both in theory and in art, includes many concepts and trends, with J. L. Borges (Argentina) and G. García Márquez (Colombia) representing the two opposite poles.

In his *Historia de la Eternidad* (1953) (quite a programme in itself) Borges asserts that to conserve and to create are inimical (*enemistados*) here but synonymous in Heaven. Here, i.e. in history, in the world, conservation means reproducing almost unchanged, while creation implies destruction, a *tabula rasa* with a new start, the arrival of the other. Heaven is eternity, where opposites become synonyms, because creation is a ceaseless recommencement, without principle or end, with no annihilation; to conserve and to create are two ways of saying the same thing: creation conserves because it reproduces the origin; conservation creates because it invents the repetition. Borges' concept of time operates within this closed dialectic.

His 'history of eternity' is anything but a chronology in which effects follow causes and consequences their antecedents. Far from recording progressions, it delves at a single point, the unheard-of moment in which in an outlying part of Buenos Aires, the mediaeval Johannes Scotus Erígena, the horizon of the pampa, and a man of today out walking meet—an encounter which might be due to 'what is known as chance', as Borges ironically writes, if time had a different consistence from that of the *unfathomable instant*. It is by accumulating on an instant that time becomes real and intelligible, the duration of past, present and future being but the sum of the elements which single out this unique instant. In one of his poems, Borges indeed writes: 'What matters the passing of time if there has been a plenitude, an ecstasy, an evening.'

However, the instant, which establishes time, is no fleeting imprint but is, on the contrary, indelible, the revelation of eternity. 'In every destiny, however long and complex, only one point really matters, the moment when a man realizes, once and for ever, who he is' (*Biografía de T. Cruz*). Time is a succession of moments, but a single moment may establish it, for, as a system, in its passage and in its breaks—in short, in its material reality, time is unseizable. Time merges with evanescence.

Borges makes time the synonym of identity, but a poor kind of identity, for eternity implies difference, disturbance, disruption. A mongrel by definition, time includes differences which, separate originally, will always bear the stigmata of their isolation. Hence the vital role of memory in Borges; for it is memory which links the differences of which identity is made, and allies the lost past with the irremediable present and the improbable future. But it is also memory which forgets and obliterates, which shows us what time is in essence: the supreme disintegration, the gaping void.

Time is a stubborn illusion. Men believe they *are* in time, but bear with them death, which abolishes time, removes them from it and spells their dissolution. Borges' characters recount, dream or imagine their own deaths so as to relive their lives, never to remodel, modify or imagine them different, but simply to fulfil and make them worth their price: death, no longer dreamt but real. Death, which stops memory, reveals Borges' conception of time: *it cannot be true, because it changes.*

In Borges' work, the constant references to universal culture (in which he does not include the Indian or black past of Latin America) complement this conception of time. Borges cannot praise Juan Moreira without recalling Platonic philosophy, nor the hanging gardens of Babylon without evoking a pink house in the suburbs, for they belong neither to past nor present and still less to the future. They have all existed at specific times but have since, through transition and change, become the ephemeral products of memory, temporary escapees from death. They survive today as what they always have been: moments of 'an impoverished eternity, without God or even a substitute, and without archetypes' (*Historia de la Eternidad*).

The work of G. García Márquez can be understood on the basis of Carpentier's dictum that Latin America is a continent where mediaeval man and modern man can still shake hands. This overlapping to a certain extent sums up the Colombian novelist's conception of time. This is somewhat paradoxical, for time is constantly present in his work but is rarely explicitly considered either by him or by his characters. His characters live wholly within time, their immersion in it being in fact the theme of his best-known novel, *Cien Años de Soledad*; but they treat it as a matter of course, like the air they breathe.

García Márquez has himself said in a recent interview that he is not interested in time as a concept—because, as a Romantic, he hates theory but loves the emotions. It is dubious, since he treats time simply as a disembodied chronology of events and characters. The time he is interested in is tufted. Multiple, simultaneous layers ceaselessly cross and fecundate: *time is qualitative*. This myth treatment, which equally rejects an arid empirical succession and the exaltation of pure subjectivity, certainly produces a spectacular effect; it demystifies, and demonstrates that myth does not necessarily lead to escapism. It is not time which is mythic, but the way in which it is treated and presented.

García Márquez' novels and short stories always tell a story that concerns time. They have, of course, real, live characters, but these always represent a certain relationship to time and a certain way of regarding time relationships. In *El Coronel no Tiene quién le Escriba*, the retired colonel Buendía is expecting his pension. The tempo of his day-to-day existence, and that of his wife—and the sale of his fighting-cock—is set by this waiting for an official communication which never arrives. Eréndida, in the story of the same name, reckons her debt towards her grandmother not in terms of service or objects, but in years.

Time separates and unites people, brings them together and sets them apart—defines them. It may create an impassable gulf between contemporaries or, on the contrary, weld together in a single present, imaginary ancestors and great-grandsons in the cradle. With García Márquez, time is full-blooded or, as an anthropologist might say, juicy.

In any case, relationship to time defines two types: those who control and impose other people's time, and those who live time at their own rhythm. The former, personified by Mister Herbert in *Cien Años de Soledad*, introduce a periodization of time from outside and decide its chronology. They settle in the village of Macondo—South American space constantly recurs in García Márquez—make some material or moral gains and depart, leaving emptiness behind but no memories, victims but no partners.

By contrast, those who do not seek to control time from outside but accept it as a natural environment are able to love, work, build—stubborn, indomitable Titans. Colonel Buendía cannot, must not, receive the letter which could switch on official time, remote, alien. The colonel depends on this; consequently, he vegetates. When, by contrast, he was a young soldier travelling up and down the country, winning and losing battles, he was very much alive. That was *his* time. This is why the 'controllers' disappear, the 'stubborn' remain *in* time even after they die, for they continue to wander about the houses, claiming attention and a voice in village affairs.

García Márquez' works are chronicles which, by 'presentifying' all times in one and by dissolving a single time into innumerable separate pieces, seek to build up a people's memory and let us know how time can be lived in Latin America. The disappearance of Macondo, at the end of *Cien Años de Soledad*, can be understood in this sense: as the end of one world, and the beginning of another, still to be created.

Eminently positive and open, time in Márquez is the necessary scenario for invention and discovery: the past is never for reminiscence but the recuperation of roots, and the edifice will be all the more solid in consequence.

IX

Two general remarks in conclusion.

None of the attitudes mentioned is the single, only authentic Latin American attitude to time. Each represents an aspect. Rather than set up one as a measure of the rest, their relations to each other and possible affinities and oppositions have been outlined.

It nevertheless seems legitimate to speak of the Latin American attitude to time, on the understanding that it includes an interrelated set of different conceptions. It definitely has a lowest common denominator, common alike to the eternal, historical, Enlightenment and Romantic outlooks. Time fully integrates all of the vital elements of Latin America and is one of its indispensable factors. Nothing can be done in Latin America without adopting an attitude towards time, without giving priority to the *conservation* or to the *transformation* of the past, the present and the future.

For, as will be obvious, time cannot be dissociated from the profound economic, political and ideological rumblings at present affecting all of Latin America, or from the fierce opposition between those who want to galvanize the present and those who want to petrify it.

In practice as in theory, in politics and in literature, time has never been so decisive or so fraught with conflict as today. Never has so much depended on time's secret: the future.

BIBLIOGRAPHY

ALBERDI, J. B. *Bases*. 1852.

Anuario Martiano. National Library of Havana, 1969, 1970, 1971, 1972.

AROSEMENA, J. *Notas para la introducción a las ciencias morales y políticas*. 1840.

BEYHAUT, G. *Europeización e imperialismo durante la segunda mitad del siglo XIX*. Montevideo, Universidad de la República, 1963.

BONDY, S. *Existe una filosofía de nuestra América?* Buenos Aires and Mexico, 1969.

DEPESTRE, R. Problema de la identidad del hombre negro en la literatura antillana. *Islas*, vol. 11, no. 1, January–August, 1969.

DORFMAN, A. *Imaginación y violencia en América Latina*. Santiago de Chile, 1970.

DUMAS, CARRERA. *La dimensión histórica en el presente de América*. Caracas, 1972.

ECHEVERRÍA, E. *El dogma socialista*. 1840.

HALPERIN DONGHI, T. *Histoire contemporaine de l'Amérique Latine*. Paris, 1970.

INGENIEROS, J. *Hacia una moral sin dogmas. Lecturas sobre Emerson y el eticismo.* Buenos Aires, 1917.

——. *La evolución de las ideas argentinas.* 2 vols. Buenos Aires, 1918–20.

Jorge Luis Borges (by several authors). Paris, 1964.

KILGORE, W. J. One America, two cultures. *Journal of Inter-American Studies*, vol. 7, no. 2, April 1965.

KULIN, K. Planos temporales y estructura de *Cien años de soledad* de G. García Márquez. *Union* (Havana), vol. 1, no. IX, March 1970.

LASTARRIA, J. V. *Investigaciones sobre la influencia social de Conquista y del sistema colonial en Chile.* 1844.

NUÑES COELHO, E. Algunas ideologías no Brasil de hoje. *Sintese Politica, Economica, Social*, third year, no. 10, April–June 1961.

O'GORMAN, E. *Fundamentos de la historia de América.* Mexico, 1942.

PEREIRA DE QUEIROZ, M. I. L'Amérique Latine: unité ou diversité? *Le Monde non chrétien*, vol. 85–6, January–June 1968.

PICON SALAS, M. *De la conquista a la independencia: tres siglos de historia cultural hispanoamericana.* Mexico, 1950.

——. *Dependencia e independencia en la historia hispano-americana.* Caracas, 1952.

PIZZARO, A. *El creacionismo de V. Huidobro y sus orígenes.* Santiago de Chile, National Library, 1969.

RAMA, C. Cultura, historia y naturaleza en América Latina. *Luso-Brazilian Review*, vol. 8, no. 1, June 1971.

RODO, J. E. *Ariel.* 1900.

ROMERO-BUJ, S. América Latina, continente del mestiso. *Mundo nuevo*, vol. 44, February 1970.

SACO, J. A. *Historia de la esclavitud.* 1831.

SARMIENTO, D. F. *Life in the Argentine Republic in the Days of the Tyrants, or Civilisation and Barbarism.* Transl. by Mrs Horace Mann. New York, Hurd & Houghton, 1868.

SIERRA, J. *La libertad.* 1880.

SOLER, R. *Historia de las ideas en Hispanoamérica.* Panama, 1960.

VILLEGAS, M. *Panorama de la filosofía iberoamericana actual.* Buenos Aires, 1963.

ZEA, L. *El pensamiento latinoamericano.* Mexico, 1965.

——. *Esquema para el estudio de las ideas en Iberoamérica.* Mexico, 1956.

Some illustrations of temporality as experienced: the soothsayer, the prophet, the guru, the leader, the futurologist

THE SOOTHSAYER: THE ANIMIST CONCEPT

Boubou Hama

To the animist, the 'object of thought' is not only that object, exposed specifically to our curiosity. It is above all its 'other dimension', its 'double' or, in the event of its being inanimate, its 'spirit', which we perceive, not in the realm of make-believe but rather with our inner eye, through our creative imagination, in which it finds concrete expression. It is in this context that the 'reality' which shapes our being and launches us on the path of our destiny is acted out.

It is principally at this level that I shall attempt to give an account of animist thought, in the fifth dimension from which it draws its inspiration through contact with its own universe, no less tangible than the world around us, which we perceive with our senses.

The animist's logic—a spiritual one—is rooted in the essence of beings and things, effecting a synthesis in which matter and spirit are blended. Animist thought is all-embracing and makes no distinction between matter and mind. For the animist, unity does not derive solely from the spiritual or material analysis of an object, but from its innermost significance, in which both physical and spiritual worlds are compounded.

Animist thought appears to reconcile the introspective philosophy of ancient India with that of the modern technological civilization of the West. It is concerned with both man's spiritual education and his material development, which is bound up with the environment that sustains him.

The animist knows that time is a continuation of time and space is but a projection of space!

The meeting-point of time and space lies hidden in the infinite, in which the animist appears to see the reflection of the world, man and man's history.

Time, space and animist thought

The leaves and blossoms on the trees close up at night and open again as day dawns.

At dusk, the worker bees return to the hive, the ants to their nests, and birds tuck their bills under their wings in grottos and in the hollows of trees or amongst the leaves.

The succession of day and night, the weather and the seasons all govern the rhythm of the life of animals, their periods of rest or activity, and affect the behavioural patterns of their multifarious ways of life.

There is not a person—man or woman—in the world who is still in the prelogical stage, not a people that does not possess its own view of life and of the world and is not concerned to fathom the purpose of life, the ultimate goal of man's destiny.

Each people has been obliged from the start to ponder the problem of its origins, of its past which issues into the present; and the latter is lived as the path towards the future.

The unity of all being has also been borne in upon the individual within the community which fosters him. This unity is founded in the creature of flesh, on his cellular life and his double, dwelling at the core of his human nature.

This unity is a past, and it is also the present which bears the individual along in the line of the general movement that animates the universe in its material and spiritual activity.

The ultimate purpose of this intemporal individual's destiny is intimately bound up with that of the doubles of the village community and that of his ancestors' doubles, on which the permanent spiritual strength of the clan or tribe is founded.

In animist thought, past and present unite in the oneness of space and time, to enter upon the future, a future which is influenced by the context of other presents which will in turn temper the conduct of the generations of mankind destined to succeed each other on earth.

Time linked to space exists on both the temporal and the intemporal planes. The latter is considered not as an idea but as another tangible state of time and space. This state gives our tales and legends a fifth dimension, that dimension which pertains to what is called, I believe, in the West the creative imagination.

Animist thought is rooted deeply in the amalgam of temporal and intemporal time. This fusion of two time-scales is the reason for the material and spiritual unity of beings and things. This coalescence of past and present, of time and space can be seen in the following strophe of the formula by which the Songhai animists begin their magic:

Ay sallü N'Debi ga,	(I appeal to N'Debi,
N'Debi sallü N'ga koy ga,	N'Debi appeals to his master,
Ay sallü bene Iyey ga,	I appeal to the 'seven highs',[1]
Ay sallü Ganda Iyey ga,	I appeal to the 'seven lows',[2]
Ay sallü Weyna huney,	I turn to the eastern horizon,
Ay sallü Weyna kaney,	I turn to the western horizon,
Ay sallü Azawa Kanbe,	I turn to the northern horizon,
Ay sallü Dendi Kambe.	I turn to the southern horizon.)

In this strophe it is N'Debi,[3] a demiurge, who is given pride of place and not his master, that is, God, whose existence the Songhai animist recognizes without, however, worshipping him in particular.

The animist, then, believes that there is on the one hand God, whom he does not name, but whose existence and primacy he recognizes, and, on the other, N'Debi, who is given the task of managing and safeguarding the material and spiritual universe. This universe is our contingent world. We subsist on its integral physical matter and on its spiritual reflection, which is perceptible by priests and people who are psychic. What interests the animist first and foremost is not God but the prime instigator, who has been given the task of administering the physical and spiritual world.

The unity of matter and spirit, the oneness of man and his double are the constant concern of the Songhai animist. The interaction of spirit and matter (even raw matter), of trees or men and their doubles, dominates and conditions all that we call Songhai magic.[4] The Songhai do not think of this magic as transcending the order of nature created by God.

While God is not immediately apparent to our senses and cannot be grasped by our reason, our intelligence, which is not of the same order as God, we sense him through the reflection of our thought upon the subject of his existence, through the mystery which surrounds his creation, from the mineral and plant world to man. God, immanent, permanent and absolute, exists outside space and time, life and matter and their essence, which is constantly evolving towards him and which he invests with his being. The Songhai animist, however, believes that God resides neither in time nor in space. He dismisses God from his thoughts, while at the same time recognizing his enduring, primordial existence, immanent but withal remote, in relation to his *art*, that is, to his immediate concerns, focused exclusively on physical and spiritual nature.

Within this nature, he prefers to appeal directly to N'Debi, who represents God within the universe of which he is part.

According to the traditions of the Songhai animist priests, N'Debi was a man who once lived on earth. In olden times, he was the elect of God, who had entrusted to him the secrets of the whole universe, the secrets of matter, its forces and spirit (when matter is pure), plants, animals, their physical bodies and doubles and the soul of man, which the Songhai animist does not name, just as he does not name God.

Although this question of the animist's conception of matter, its spirit and energies do not pertain directly to our subject, it is, I think, worth dwelling upon briefly, as well as on the question of plants, their inner life and their doubles (closely akin to those of animals, albeit not identical).

The animist regards the spiritual element in the order of nature as tangible, just as matter is tangible. The same is true of the doubles of trees, of animals and of men, all palpable spiritual entities as perceptible to those who are psychic as the fodder we feed to our horses, the wood from which we fashion a chair or table, or the tender flesh of the animal of which we make a delicious meal.

Matter, its energies and spirit, trees, their wood or bark, animals, their skin, bones and flesh, and man himself, his flesh and his double, do not transcend the order of nature, a perfect whole at the centre of which N'Debi, the demiurge appointed to watch over the universe (and to rule it, observing the strictest rites), sits enthroned.

This demiurge was once the prime mover. What he taught the first initiates constitutes the sum of knowledge with which, on behalf of God, he has endowed our common humanity. At his death, the initiates scattered throughout the world, establishing schools and centres for study and initiation, of which our castes today are but pale reflections.

The Songhai animist believes that the men of old (*don borey*) were closer than we are to nature, upon which they exercised limitless power, thanks to N'Debi's teaching.

This demiurge lives on. In the intermediary world, he dwells in that intemporal time in which the spirit of matter and the doubles of plants, animals and men all forgather.

This intemporal time informs inorganic bodies as well as the living or dead matter of which plants, animals and human beings are formed.

By the practice of sympathetic magic, it is possible to make contact with a mineral, plant, animal or man if we possess a constituent element of its physical body—a splinter, a leaf, a strip of wood, skin, a piece of fur, a bone, or even its reflection in a mirror or in a pool, or its shadow upon the ground.

Thus the Songhai animist holds that once one has a part of some Whole, one possesses that Whole. Wherever the Whole may be, it is at the animist's disposal, before his very eyes, within reach of his hand. This is why he takes particular care to bury his nails and his hair after he has cut them, lest they fall into the hands of his enemy.

The animist distinguishes between the physical and spiritual universe, which is the realm of N'Debi, and God, who is distinct from this contingent world, while at the same time exercising an influence over the beings and things, the spirits and divinities which dwell in it.

The animist, then, believes that, outside and beyond the tangible world around us, there are two other worlds, as tangible as the world we live in: the world of God and of faith, which is inaccessible to his powers of investigation, and that of N'Debi, which he apprehends by his own powers and by the application of his science.

The strophe reproduced earlier puts these two spheres, of which N'Debi is the common element, clearly into context. N'Debi has charge of beings and things, the matter of which they are constituted and their doubles or spirits, but not of their souls, which pertain to a different spiritual sphere, whose existence the animist senses and is aware of, but whose immanent presence is inaccessible to his senses and to his powers of investigation or knowledge.

Animist practice leads to the threshold of this existence, which is very close to that of God, whom the animist does not seek to elucidate, although he testifies to His timeless presence from the beginning.

When a double plagues the members of a Songhai family or the inhabitants of a Songhai village, the *Hirow* is brought in to kill it in the form of a chicken, a lizard or some other animal. Once the double is killed, however, it does not rejoin those of the community of ancestors. It simply vanishes, reduced to an essence (no doubt the soul) which cannot be perceived by our senses. Thus the soul appears to belong to a spiritual sphere which brings it close to the faith that leads on to God.

The animist turns to God only in extremity, when danger is close at hand. He addresses Him in very short formulae because of the imminence of the danger, which allows him no time to request N'Debi, through the normal channels of the order of nature, to intercede for him.

The regular formulae are rather lengthy and are called *jindi*, which in the Zarma and Songhai tongues means 'distance'.[5] The short formulae are known as *sauci*; these are commands given to halt, check or interrupt a movement initiated by a being, whether endowed with intelligence or not, or by some object or thing in the process of acting or moving.

For example, to subdue the lion, crouching ready to spring upon a man or upon his prey, the animist priest levels a forefinger in the beast's direction and declaims:

Kotombi kani,	(Kotombi[6] is couched,
A hinjey kani,	His teeth are couched,
Kotombi kani,	Kotombi is couched,
A ce mosey kani,	His claws are sheathed,
Kotombi kani,	Kotombi is couched,
A sounfa kani,	His tail is couched,
Kotombi kani,	Kotombi is couched,
A hanney kani.	His mane lies flat.)

Similarly, in order to divert this lion's wrath to another person or to an enemy whom he wishes to destroy, the priest will say:

Sa dagar,[7]	(Sa (the lion) is growling,
Zamaa ni ga dagar,	since you're growling,
Dey ni nakoy dagar	go and growl
Filanaa bon!	at *X* ...)[8]

In view of the content of the latter part of this account, I feel that it is important to define even more clearly the categories from which the time estimations used by animists are taken.

Time, the instant and being

The animist makes a clear distinction between his physical body and the spirit which invests his double. He recognizes the existence of the physical body, which is destroyed at death and is vacated by the breath that sustained and gave life to the being that dwelt in that body. This physical body is considered as being made of earth, to which it returns once its life on earth has come to a close. The animist expresses this truth when he says: '*Adamu-Ize ya laabu no*' (Man, son of Adam, is of earth, and to earth he returns).

But over and above this physical being that lives and dies, and his corporal self that returns to dust after death, the animist believes firmly in the existence of the double that dwells in his body (*gaa ham*), in his innermost being. He believes that the double perpetuates his spiritual existence after death. Little enough separates him from the beyond, from the invisible world peopled by the shades of his ancestors, the genii[9] and immortal deities who are visible only to mediums and to others who are psychic.

Man is a being composed essentially of a physical body and a double, which is external to man's life (*hundi*) and which departs by way of his nostrils at death. The animist distinguishes this *hundi* from the cellular life which is to be perceived in fresh meat, in the bark of a living plant, in the vivid colours of its blossoms and its verdant foliage.

It is his life on earth, bound up with the endless life of his double, with cellular life and with the ephemeral life of certain insects, that instils in man a sense of the duration of life, and makes him keenly aware of the destiny of those things and beings whose death is also a renewal of the life by which the species is perpetuated.

In relation to the life-span of certain members of the vegetable and animal kingdoms and to the enduring quality of stone or of the mountains that symbolize eternity for us, our terrestrial existence is brief indeed.

The animist believes that within the human being there is his subtle body[10] (his double), in which latter his existence will be perpetuated among his ancestors' doubles, in their community which is organized in the same way as earthly society, its tangible replica on earth, where we live.

There is, then, our terrestrial life and that of our ancestors' doubles, that of the society which sustains us and that of the doubles who are its spiritual foundation, a foundation linked to the life of the ancients or of the prime mover.

Earthly existence is perceptible to everyone through the senses. It is in terms of its duration that the animist measures or apprehends material or temporal time.

One of the yardsticks by which he measures such time is the life-span of certain insects. Because of the briefness of their existence, ephemera are looked upon as the breath of death, *bu haw*. The briefest instant is called *moy kilow* in the Songhai dialect spoken in Tera. In the Songhai and Zarma tongues, the time between opening and closing one's eyes is: *moy hay ka moy dabu*, that is, the twinkling of an eye.

When a child is sent on an urgent errand, he is told to go quickly and 'be back before my spittle dries on the ground' (*te tambatamba, koy ka za ay tufaa mana koogu*).

Similarly, the animist measures time in terms of a man's step: *ce medaara* (Songhai and Zarma), or *tabande* (Fulani). For example, he may say: 'You'll not take a single step before you've paid me your debt', or again: 'Another step, and your fate is out of my hands.'

The animist is also conscious of time as the regular interval between inhalation and exhalation. He believes that this time-span is the space that marks the rhythm of our breathing, which is bound up with our life, with one of its entire cycles, the length of human life, which is co-extensive with the length of man's years on earth. This span (his age) is commonly referred to as *hundi me fo*, meaning a whole life, a full life, a life filled to the brim.

This life-span is divided into the following phases: the period of gestation (*za zankaa go nia gundaa ra*); birth (*atciria*, literally, the baby being born); the period during which the baby is breast-fed (*nanindu*, the baby at the breast); the period of weaning (*ka uwaa* or *kosorow*, the weaned baby); the period of growth (the baby boy grows up to be an *arwasu* and the baby girl a *hondio*); puberty; maturity; old age; death, which marks the term of our earthly life.

The life of the old man (patrilinear descent) or of the old woman (matrilinear descent) perpetuates or rather maintains among the living not only the 'image' of the ancients but above all their 'memory'. In the older people is vested the *spiritual power* of the ancients, in the form of some sacred object or other, which is handed down from generation to generation to the oldest member of the clan or tribe, the community, the village or region.

Space, the environment, the factors and phenomena that govern men's lives and mental attitudes—all these impinge upon time: the day's duration and the different positions of the sun during the day; night's duration and its prevailing calm. The Songhai and Zarma define the different moments of the day by reference to the apparent course of the sun, using active verbs which can only be paraphrased in English: *bya* ('to set out early in the morning'); *zariney* ('to set out at midday'); *weynaa* ('to set out in the afternoon'); *almariney* ('to set out at nightfall'); *hanna* ('to set out at midnight'); *alfararey* ('to set out at four o'clock in the morning').

Time is marked by the succession of the seasons and by man's activities during these seasons, particularly the rainy season, during which he raises food crops that can sustain life. In the Sudanese savannah, a man's age is generally counted by the rainy seasons. To express the fact that a man is old, it is commonly said that he has drunk a great deal of water, meaning that he has lived through many rainy seasons.

Deaths and births are related to events that make a lasting impression on the peoples of a particular area or an entire country; for example, the year of a certain famine, a war or an epidemic, the arrival of some great personage, the coming of the whites, the death of some great chief, and so on.

The animist distinguishes between material time and intemporal time, while remaining fully aware of their interfusion which is bound up with the spirit of matter, the motions of our corporal self or the tribulations of our double.

These motions dominate our human nature, which is either patient or impatient. Patience (which instils in us calm) and impatience (which breeds unease), urbanity and incivility, courage and cowardice, generosity and egoism, wisdom, intelligence and cruelty, rectitude and guile—all are governed by the motions of our innermost being, which either experiences the impact of its environment or acts upon it to change or transform it in accordance with our wishes.

Expectation has as its theatre of operations our innermost being which it involves in the motion and in the habit of carrying out or undergoing this motion which moulds our individual characters through the permanent values of the society that helps or sustains us.

In this sphere, the animist's inner life is dominated by the past, which he actualizes in the present—and this is itself like the future. The future rests on the foundation of the past and present, and benefits from their lessons, as though illuminated by the light of other contexts which can influence the course of the future.

The 'duration of time' and the 'course of time', whether rectilinear or meandering in and out of the labyrinth of events, are of a piece and form a whole in which past, present and future interfuse in and through physical duration, which has been, is and will be the time that elapses while we wait for a promise to be kept, watch some passing scene, wait for a metal to melt, our mail to arrive, or our daily task to be finished (time measured in hours), or recover from the grief caused by some bad news, or from some misfortune that has overtaken us.

The animist knows that all this is but an instant of that motionless time in which we all live in serenity. And this naturally raises the question of intemporal time as it affects social life and the practice of animist art or science.

This notion of time is implied in all magic formulae. It ushers in the action of the priest who seeks to ward off some danger, whether imminent or remote. It is a sort of spatio-temporal syneresis extending not only chronologically but spatially too, which is appropriated by the sorcerer, by the seven heavens above and the seven heavens below, starting at the horizons to the east, the west, the north and the south.

This appropriation brings together past, present and future, which are bound up with N'Debi. The beings and things dwelling in this fifth dimension are comparable to a chemical paste that the sorcerer kneads into the shape dictated by his formulae, by the creative word preserved in the language of the genii, gods and men of old.

All is, and all remains—such seems to be the message enshrined in the animist's philosophy of life.[11]

Time, whether temporal or intemporal, derives from this conception, through which the animist resolutely affirms that man is rooted in the universe and its spirit, its matter and the energies released by this matter, in which all living beings bathe, both their quivering bodies and their doubles, whether intelligent or not, whether independent or governed by the forces or spirits of nature, in pursuit of equilibrium or the Good or in pursuit of disequilibrium, which generates sickness or causes the deterioration of our health—that is, Evil, which corrodes matter or the social order.

Time and its spatio-temporal reality

The animist has a comprehensive view of the universe, which embraces the sum of all the beings and things upon which it projects its spirit, entering into them or spiritualizing them. He sees everything as resting on a spiritual basis, which supports and sustains the spirit of minerals, the doubles of plants and those of animals and men, these latter being constantly 'tuned in' to the past, which is brought into being whenever the animist fashions some artefact or accomplishes some act, even some common, everyday act.

Let us observe the animist as he practises his art or applies his science.

In the introductory strophe quoted at the beginning of this study, the story of creation takes place in absolute, 'timeless' time, which was and is, and which the animist priest 'appropriates' by uttering the strophe, thereby endowing it with an existence of its own, an existence achieved in the present.

At the same level, this spiritual aspect of creation is given concrete expression in objects on which the material and spiritual foundation of this same creation rests.

The site of the workshop, field, or village, the boundaries of the region or country are, first, established by means of the guide marks which demarcate them. In the middle of the site, the animist priest summoned to perform the rite then buries a piece of iron dross or, failing this, a fragment of old *canari* (pottery) taken from an abandoned village.[12] This iron dross or fragment of *canari* represents the heavens' seven 'highs' and seven 'lows'.

To represent the spatial horizons, the animist places a piece of iron dross or a fragment of *canari* at the east, the west, the north and the south of the chosen site, regardless of its dimensions. Not until these operations have been performed is an animal sacrificed and its blood offered to the spirits of the place in which it is desired to set up the business and to the spirits of the ancestors, to the genii and gods that dwell in the tall trees and mountains and in the grottoes surrounding the site.

In animist thinking, the past is constantly compounded with the present. The introductory strophe of the Songhai incantation is followed by these lines:

It is not from my mouth,
but rather from the mouth of A
who gave it to B
who gave it to C
who gave it to D
who gave it to E
who gave it to F
who gave it to me.
May mine be better
in my mouth
than in that of the ancients.

After this invocation of the ancients, the sorcerer comes to the main part of the incantation whose effect is to be focused on a specific object, on some evil which he intends to root out. This main part of the incantation specifies subsidiary obstacles which he eliminates one by one before tackling the chief obstacle. In the following case, for example, the priest has set himself the task of curing a migraine:

I

Bisimilahi! *Bisimilahi*! *Bisimilahi*![13]
(In the name of God! In the name of God! In the name of God!)
I call upon N'Debi
N'Debi calls upon his Master
I appeal to the 'seven highs'
I appeal to the 'seven lows'
I turn to the eastern horizon

I turn to the western horizon
I turn to the northern horizon
I turn to the southern horizon.

II

It[14] is not from my mouth
It is from the mouth of A
who gave it to B
who gave it to C
who gave it to D
who gave it to E
who gave it to F.
It was F who gave it to me.
May mine be better in my mouth
than in the mouth of the ancients.

III

It is N'Debi's she-camel
who has entered the bush.
He has called out the bush people[15]
He has called out the village people[16]
They sought her,
they did not find her.
He told me to seek her.
I sought her,
I found her.
He told me to kill her.
I killed her.
He told me to skin her,
He told me to take the rib.
He told me to take the hoof.
I said I did not want the hoof.
He told me to take what meat I wished.
I stretched out my arm,
I took the head.[17]
Just as the head of N'Debi's she-camel
did not prevent me from taking it,
so thy headache, *X*, cannot prevent me from removing it.

IV

If the pain is in thy head,
I have removed the pain,
I have exorcized it.
If the pain is in thy nostrils,
I have removed the pain,

I have exorcized it.
If the pain is in thy eye-sockets,
I have removed the pain,
I have exorcized it.
If the pain is in thy ear-holes,
I have removed the pain.
I have exorcized it.[18]
Nought but my hand,
But N'Debi's hand,
For ever and ever, without end.[19]

This formula, which places the action at the beginning of time, fuses in a single whole both past and present, the latter being but a section of the future which protracts it in intemporal time; this intemporal time, despite appearances which make its different moments or aspects perceptible to our senses, is motionless.

Continuous time appears to be fragmented at the level of a number of esoteric figures which symbolize it[20]: 3 = the number for man; 4 = the number for woman; 7 = the number for the couple, God, and the days of the week; 9 = the number which is the origin of novenas, and the multiples 99,999, etc.; 40 = the time after which the sacrifice to the memory of the dead is made.

The magic paste, which enables the animist to vanish when danger threatens, and the paste which makes his body as hard as stone when struck by spears or bullets, is taken by men once a day for three days, and by women in four doses, once a day.

The power of these pastes, which are taken for a specific period of time, is generally delayed, and their effects are not felt till later, depending on the circumstances in which they are taken. For example, the paste which causes one to disappear is taken just as the person sits down on a millet stem placed across the mouth of a deep well. As soon as the person touches the stem, it breaks in half, but instead of falling into the well, he is miraculously wafted away to some remote spot in the surrounding countryside.

Thus, whenever danger threatens (in the shape of a wild beast's attack, an enemy's attack, a car accident) whoever has taken this paste vanishes under the effect of his emotional shock.

Each clan having a command of the skills in which iron is used (smiths and jewellers), water (Sorko, Somono and Bozo[21]), the earth (the priest of earth, masons), wood (woodsmen), or animals (butchers, hunters), and so forth, claims to possess one or more such powers in connexion with their skill or with the materials used in it.

It is such factors that characterize the animist's 'mentality', the underlying reasons that govern his attitude to life; and it is due to them that these reasons are uniquely his own, and are subject to no other logic.

The animist justifies himself in these terms:

If there were no controlling power to regulate the movements of men and stars, we should live in such chaos that one day we should see the sun rise in the West and set in the South, or the tall trees of the virgin forest sprout up thickly on the surface of the icy wastes of the North Pole. This tremendous force, which appears to be doubted by certain wise men, is not in the least doubted in Black Africa. The African believes that God exists: the Bambara call him *Massa Dabali*, that is, 'King not created' or *Fembe Damba*, that is, 'Creator of all things'. But he is so transcendent a being that only words composed in accordance with set ritual can reach him. Hence the prayers, the dances and the great fires lit during the ceremonies that are held everywhere in Africa to celebrate important events. An old adage has it that: 'One must pray unceasingly and beseech the Supreme for his assistance on all occasions.'[22]

Truth, as the animist sees it, is based on its own reality, the essence of which is the intelligent double, active and responsible.

To the Songhai animist, man is first and foremost his corporeal self, a being of flesh who breathes and labours, who thinks and ponders the pros and cons of his actions. We know him as our fellow man, the neighbour we meet at the village well, the pond, or the river bank at the market or the local store, and with whom we converse around a table or in the village square where discussions take place.

By day, the man works in his field, in his forge, his jeweller's or weaver's workshop, at his tradesman's stall, or on a building site, toiling for his livelihood. He lives his life before our eyes. We see him reading, writing, thinking, carrying out some specific job. At the end of his day he returns, weary, to his home. After his evening meal, he stretches out on his rush mat or his bed and falls asleep. This, the animist believes, is the moment selected by his double to retrace his steps, haunting the places frequented by that man by day, re-enacting all that he has consciously done during his daytime life.

It is during these wanderings that the double falls foul of the forces of good and of evil, the evil genii and sorcerers that devour men's doubles, called *cerko* by the Songhai and Zarma.

The individual's personality dwells in this double. The Songhai say of a man that his *bya*, his double, is heavy or light, meaning that his personality is strong or weak.

The animist's art or science is focused upon the physical body, on the spirit or its double, on the human nature which they sustain and, what is more crucial, upon its double, in order to enable it to reach ever loftier spiritual spheres, the spheres inhabited by the foremost spirits or powerful genii which may choose to attach themselves to a human being. The Songhai do not say of a man that he is a genius but rather that he possesses a genius who assists him. This genius is to be identified neither with the man himself nor with his double, both being wholly independent in their intimate fellowship. The amulets worn by the animist and the magical pastes which he swallows are intended to fortify both his body and his double and to make the latter capable of mastering the forces or spirits of the intermediary world of which our own is but the material representation. The most powerful, the most charismatic priest is the priest who achieves a state of positive consciousness merging into and becoming one with his double.

At a certain stage in his initiation, or upon reaching a certain degree of wisdom, the animist succeeds, thanks to his double's efforts, together with the external powers, in spiritualizing his corporeal self, his human nature, which has become fused with that of his double. In this state, the full initiate, the grand master, the *korte-koynii* and the *zimaa* enter another dimension of matter and being in which 'everything', now sublimated, becomes possible, and can be achieved in response to the slightest desire of these luminaries of animist religion. In this dimension of matter and being, time fuses with space and man becomes his essential self, and is thus able to attain his own enduring reality.

The question of the spirit and the double is bound up with the very origins of matter and being, and these are connected with the essence of the physical and spiritual universe in its totality. Being a universalist, the animist reifies this totality, effects a spiritual synthesis which confers upon his art and his science their eminently spiritual character. The animist does not conclude that matter is void or that being is not invested by its double. To him, all is *One*, yet different according to the way in which we see the single All, concealed by its various guises, which distort the inner reality of beings and things. This being so, the animist believes that everything endures, even when it cannot be perceived by the senses.

Material time exists in relation to us and to the things which we perceive during the different periods of the day and night, the changing course of the seasons, events or phenomena the effects of which impinge upon us, the duration of their effects or of the cessation of their activity, their intensity or their violence, the momentary pleasure we have in them or the displeasure they cause us, through the birth and death of beings and things, through our own existence, our joys and griefs, successes and failures during our life on earth.

However, this material time is closely related to intemporal time, which it marks with its regular moments and divisions: the succession of day following night, the movement of the heavenly bodies gravitating around the sun, that of the stars illuminating the pure African nights with their fairy-like twinkling.

In this context, the convergence of temporal and intemporal time becomes a spiritual reality; the animist refrains from seeing God here—God, who is other than this reality, but who nevertheless animates it with His health-giving breath, with the spirit of N'Debi, at the centre of that unchanging convergence where the destiny of beings and things is acted out.

Nor is this consummation of time named by the animist; he is simply aware that it exists.

In this, he is acting as a believer. He does not deny God. He leads us to Him in this absolute time, leads us to the threshold of the soul and its creator, to faith by which we reach the godhead.

The double is like our image perceived through the untroubled surface of a pool of water, or reflected in a mirror. The past is hidden in it, and it expresses the past by prolonging it into the present. It exists in a state of matter and being which lends itself to all possible transformations. It is by using this resource that a man who has consumed a certain paste can vanish, traversing vast distances, that his double can be transformed into a chicken or a lizard which the *hirow* or *cerko* (the double-devouring sorcerer) then kills, and that his inner being, the spiritualized object, can be metamorphosed into a stone, a grain of millet seed or water, and vanish without a trace. This power of metamorphosis is not the product of a single human generation. It is a collective power, giving material expression to the common striving of the clan or the tribe, to the sum of their efforts materialized in an object which can be transmitted by the elder of this clan or of that tribe.

This naturally leads us on to the study of the clan's real power. To the animist, this is not a mere matter of social pressure.

The clan's real power lies in its spirit or collective soul, on which its stability rests. This power, which, the Songhai animists believe, is vested in a particular object, exists in each family, in the safe-keeping of the eldest member, who hands it on in this material form to his successor. The same is true of the clan and the tribe. The object itself may be a golden ball which is kept in a *tobal* (war drum) together with elements taken from the lion, the symbol of reckless courage, or from other animals, such as the elephant or the panther. It may be enclosed in a box, in a piece of *canari* or in any other similar object. It may be a pointed steel blade (for example, the *lola*, among the Zarma and Songhai). The *Sorko* of ancient Gao had an idol in the form of a large fish with a ring in its mouth. Among ironsmiths, it is represented by a mythical forge that glows ruddily at night to show its anger (for example, among the Doutchi Baouras in the Republic of the Niger). This power can give life to a human corpse and enable it to designate the dead chief's successor. Among the Cerko, Songhai and Zarma, it takes the form of gelatinous, shell-less eggs. Among the Sonianke, it is found in the guise of an exquisitely worked gold, silver or copper chain. Among the Songhai, it sometimes occurs in the form of a standing stone called a *jingara*, meaning the mosque.

According to Songhai legend, the great mosque of Gao was surrounded by forty stones representing the forty genii that guarded the mosque. In 1591, when Gao was taken by the Moroccan troops, each of the Songhai clans took possession of one of these stones, which symbolized for them the mosque and the past grandeur of the Songhai empire. Even now it is the object of a cult in many villages in the Tera region (Niger). Among the Gourmanche of Doumba (Tera, Niger), this power was, until the advent of the French, incarnated in a lion. Elsewhere it is represented by: a mythical serpent (the pool of Oslo, the pool of Kokoro, the river Dargol, the Sirba (Tera, Niger)); a crocodile (*Tola*), among the Kourtey of old. (This explains why the Fulani still refer to the Kourtey as *Tolabes*, that is, the crocodile people. The worship of *Tola* was at the origin of that of the sacred cayman, which was widespread in many Gourmanche villages before the colonial era and which, the inhabitants held, represented their deceased ancestors); a stick, with a vulture's head carved in the wood (found among certain Sonianke clans in Wangarba, Tera, Niger).

Occasionally, too, it took the form of a gold cooking-pot, a mythical net, a hammer and tongs, an axe, a stone (frequently a stone axe), a calabash, a cap, a sabre, and so forth, in all of which the collective soul of the clan or tribe might be contained.

Other tribes worshipped this power in the guise of a gold figurine representing a chained man (in the present-day district of Dargol, in Tera, Niger) or of the masks belonging to the family, clan or tribe, which served as its receptacles (among the Dogon and Bambara tribes in Mali).

The African animist is constantly linked to his past through the medium of these objects, either individually or in association with other objects and with his traditions. This brings me naturally to the question of the role of temporal and intemporal power in traditional African societies.

These two powers are not identical, although they may co-exist in a single person or in a single object.

When the chief of the clan or of the tribe or the head of the family died, it was always the elder who was appointed to succeed him, unless some prohibition (a physical defect or infirmity, an impediment, or some attendant ill-luck divined by the soothsayers) stood in the way. So long as the object symbolizing the power had not been placed in his keeping, he could not be considered the real chief. The prosperity of the family, the clan or the tribe was—and could be—enshrined only in this object, which alone represents the real power, the enduring essence of the human entity concerned.

Because they misunderstood African society, the government authorities frequently failed to take this crucial fact into account when appointing chiefs. During the colonial era, there were as a result temporal chiefs who were rejected by their peoples because they did not possess the spiritual power, the object in which it was enshrined, the *real power*, temporal and spiritual power, which would have established them as the true, incontestable chieftains. When the chief appointed failed to fulfil this essential condition, his people, if they were able, appointed a spiritual chief to assist him in his task.

At Wanzarba (Tera, Niger), where descent is matrilinear, the French authorities appointed a man to be village chief, so as to bring this village into conformity with the other Songhai villages. In their place of worship, however, the Sonianke naturally retained their priestess, the *Kassey*, and she continues to assume the responsibilities vested in her by the spiritual power.

In this clan, the spiritual power is passed on primarily by milk, although it is acknowledged that blood ties strengthen this link. Among the Cerko, however, the power is transmitted solely by way of milk.

On the spiritual plane, there exists an antagonism between the Cerko and the Sonianke. The former are the opponents of society and the latter are its defenders. Here we see the old feud between matriarchy and patriarchy continuing at the spiritual level.

The confrontation between the Sonianke and the Cerko takes place at night. The Cerko represent the past and the Sonianke the present. The Cerko seek to rule over the night. The Sonianke, who are lords of the day, attempt to wrest from them the night in which the struggle between the two clans is joined.

At Doutchi, neither the chiefs nor the princes are permitted to see the *saraounia* of Lougou, the priestess of the earth and its first owner.

In the same country, the *Baoura*, who is considered to be the maternal uncle of the Doutchi chiefs, never sets eyes upon them. After the advent of the French, the difficulty which this created was to some extent obviated by hanging up a black loin-cloth between the Doutchi chief and the Baoura whenever they appeared together before the commandant, in order to prevent a possibly tragic confrontation from taking place. There is no doubt that the black loin-cloth represented night. In Benin, there was a king of the day and a king of the night.

Is not the African animist revealing here, as though for our entertainment, an aspect of his life in which the past rises up against the present in order to oblige it to heed its lessons?

In the examples which follow, this conflict within the clan or tribe is fortunately less pronounced. Within these human communities, the spiritual power is transmitted intact, without violent confrontations.

Each link of the Sonianke's chain represents an ancestor. The chain as a whole symbolizes his ancestral line, from the primordial ancestor down to himself. I myself have seen such chains, over one metre long, made of gold, silver or copper or some substance indistinguishable from these metals. To the Sonianke, this chain represents the clan's succession of ancestors, the clan's past being embodied in the chain, which is 'regurgitated' by the Sonianke and transmitted to his successor at the moment of death.

These chains have the glitter of metal in another state of matter. They are not deposited in the Sonianke's stomach in the same way as the foods which we consume. They issue forth from his innermost being. The Sonianke transmits his chain to his son, to a blood relation, but never to anyone who is not a member of the clan and has no right to it. The chain is swallowed by its free end at the moment at which the dying Sonianke regurgitates it. He dies immediately after delivering it up to the person who is to carry on from him.

These remarks may shock the reader. They do, however, illustrate clearly the mental process followed by the animist, which is closely bound up with his conception of man and life, and is further evidence of his belief that spiritual forces exist no less really than the material world which we perceive with our senses.

The hallmark of the black African, that which stamps his behaviour, controls his conduct and distinguishes him from the Westerner—the product of a civilization on the verge of becoming dehumanized—is his depth. The animist possesses a philosophy of life which unites past and present in some enduring object that defies the action of time, which is constantly delayed in its progress towards the future.

Immutable time is perhaps the seemingly motionless earth and the sky that girdles it. If we disregard the stars, the sun and its harsh light, the life that teems upon our planet, the myriads of insects, plants and their wonderful blossoms, animals and men, and the phenomena that throw our earth and its atmosphere into turmoil (winds, hurricanes, stormy seas, volcanic eruptions, earthquakes, and so forth), the universe appears to be set firm in its origins, in the very cradle of life and time.

Nevertheless, we sense invisible presences about us, an infinity of sounds in nature's austere silence. All things have their place and are interrelated so as to form a whole. Undoubtedly, there is no vacuum in the universe. A story told by the late Quézzin Coulibaly, the great leader of the Rassemblement Démocratique Africain, is of interest in this connexion.

During the colonial era, circumstances brought together the elderly African, Losso, and a French doctor by the name of Kremer.

One night, the Frenchman sought to hypnotize Losso by the light of the paraffin lamp. Losso met the doctor's steady gaze with a somewhat ironic smile, and said: 'Doctor, I know what you're up to. But in this sphere we Africans are more than a match for you.'

Losso took Kremer's lamp and blew it out. The two men were plunged into total darkness. Then Losso raised his forefinger and breathed on it; immediately it radiated a dazzling light. By the light of this human torch, the two friends continued talking late into the night.

At the end of their amicable discussion, Losso observed: 'To make light, you needed: a lamp, a lamp-glass, paraffin, a wick, a match and a matchbox. To light the lamp, you then had to strike your match on the matchbox and set light to the wick, so as to obtain a steadier flame. But

I can make a better light without any cumbersome apparatus, simply by raising my forefinger and breathing on it.'

And he added: 'You whites possess the *power of matter*, but we blacks possess the *power of the void.*'

The animist always starts with the spirit. It governs everything—all his actions, and all that he claims to possess: his village, his plot of land, his hut, his country, his earth, his craft and tools and his membership of a dynasty or a caste …

The plot of land, the village, the forge, the weaver's workshop, the field, his country are his by virtue of certain immutable material and spiritual principles. The farmer, who is lord of the earth he farms, makes it barren by cursing it and by striking it with the flat of his right hand, for he is the spirit of the earth.

The ironsmith, the *Sorko*, the butcher—to mention only these—all perform public ceremonies in order to prove that they have not lost the secrets of their craft or trade and the powers attached thereto.

During public ceremonies performed in the village square, the Sonianke dances to the rhythm of the tomtoms and regurgitates his chain.

In order to vindicate his right to his position, the *griot* recites from memory the history of the peoples of his region or that of the traditional chiefs in whose court he resides.

Man is the reflection of his own purpose in the universe, a burning field of energy, good or bad, which draws upon him good luck or evil, the source of that disequilibrium which generates chaos and disease.

In his innermost thoughts, the animist is constantly preoccupied by his origins, by the essence of things and their harmony, which sustain him in his daily tasks. He attaches more value to the spirit than to matter, and sets great store by the past, in which the core of his being is rooted and which forms the mainstay of his everyday life—a life not opposed to change—that is, to progress, by which is meant neither drug-taking, nor the unrest of today's consumer society, nor its doubtful comforts, nor its automation, nor its marvellous technological achievements which eliminate the human element and can drive men mad or make robots of them.

The animist's synthetizing mental approach spiritualizes the subject of science. The animist makes man once again the 'proper study of mankind'—and with man, nature, which is invested with his vital spirit, religion, man's need to believe in something and first and foremost in himself, and in the universe around him, a universe from which he does not exclude God or divinities.

The animist leads us to the threshold of faith and God. The African kingdoms and empires of the Middle Ages (Ghana, Mali, Gao), on the shores of the Senegal and Niger Rivers, in Bornou and in Chad arose from the encounter between his all-embracing philosophy of life and the world of Islam.

The past is the leaven which gives the present its dynamism and our personality its pride.

Civilization is not the exclusive property of the West. Western civilization has aroused the interest of the whole world essentially because, at the technical level, its human achievements are admirable and, in my view, irreplaceable. It must, however, understand that mankind may not wish to be confined within science as it has developed in the West. Now is perhaps the time for Western civilization to accept the idea that other modes of thought, including African animist thought, can infuse into it the vitalizing spirit which will enable it to pursue its onward march.

Our planet functions as a single, cosmic unit, and the risk of death as a result of radioactivity is one to which we—the human race—and the planets and animals with which we share the earth, are all exposed.

The deterioration of our oceans, war, disease, hunger and pollution all point to the unity of the planet in the diversity of continents and of their peoples, which are growing ever closer to one another.

All life in the modern world is sustained by the sum total of the crops we raise and of the raw materials available for all the members of the human community. What Black Africa, a harmonious synthesis, has to offer mankind is its cosmic civilization, which is in the process of engendering a new mode of life, one that is already potentially present in its arts, its cultures and its thinking.

Our contribution to Western civilization is a *dynamic slackening of pace*, which may enable it to rediscover its soul, and the hope of creating a new, *whole man*, firmly anchored in both his spiritual and his material nature.

Animist thought, which is synthetic, generalizes the deeper significance of the object; it does not seek to analyse it. Its approach is an all-embracing one, which can grasp the nature of the object as a whole.

The life of the individual, its origins and day-by-day unfolding, indeed, its very briefness, presupposes the uniformity of time and the confrontation of this uniformity with that of life as reflected in the enduring existence of the different species, so that this existence appears to be the ultimate purpose of life, to which each individual merely makes his contribution.

The animist sees man as being in a transitory state on this earth. To wish somebody long life, the Zarma and Songhai say: 'May God grant you an extra two days', that is, in relation to infinite time, a brief respite on earth.

The animist is aware of the tragic nature of this time. He sees it in the brevity of his own life-span, in nature's cycles and in the rites by which nature is renewed or restored to her primordial purity. The animist knows that physical time, in which his own life runs its course, is the measure of his emotions, of his desires, whether satisfied or not, of his impatience when he hates and his pleasure when he loves. He experiences temporal time as but an external factor linked with his corporeal and subjective being, his terrestrial life as related to that of other beings and things.

The perpetuation of the species suggests to him another form of time, to which he seeks to give material and spiritual expression in concrete objects that can be handed down from one generation to another and that can embody or enshrine in temporal time those social facts and historical events which originated in the past of his ancestors.

The animist associates temporal and intemporal time in all his actions, in the social order which sustains him and in the presentation or representation of his history. To him, social facts and historical events must both, if they are to belong to the order of nature, accord with that reality which laughs at our passions, our self-seeking ways, our bloody wars and the natural calamities sometimes visited upon our earth.

For the animist, such time presupposes that infinity in which man plumbs the innermost depths of his being and of the universe's apparent emptiness, that other dimension of man and matter to whose mystery he alone still possesses the key.

Temporal time issues out of intemporal time, setting its seal upon those events that stand out as landmarks of history.

Time and history

This section will deal with temporal time, with the instant and with spatio-temporal time, particularly in so far as these affect the animist's life in society, in which he acts out his *reality*, the projection of a distant past continued in society by *the elder*, who carries on the true tradition, as many adages show: The chief's word is the mark upon the stone; True knowledge is inherited (a Hausa proverb); However the doe may bound, her fawn will not lose its track; Water does not change its course; May this ancestors' milk be at peace in thy belly; May God give thee the good luck of the ancients; Water never kills its offspring.

The elder represents the truth and wisdom of the men of old. He is the past and the present, and he will be the future. He is the whole truth revealed by N'Debi, in close relationship with the universe's double.

Tradition—embracing art, culture and history—has been and is still this truth, shining like 'the sun, which cannot be screened with the palm of one's hand'.

In view of all this, one might venture to suggest that the animist has no history, since in his view time does not exist—all is correspondence, assimilation, even repetition, reduction to type.

Today, all paths forward are a retracing of our steps.

This idea, however, is contradicted by the being and the very actions of the person who takes the place of the prime mover. Truth, which is recalled in its pristine purity, is continued in series of actions or exemplary lives which could not follow each other in many different contexts without some discrepancy. History is made up of these necessary adjustments, which mark its different stages.

The animist is sustained less by history than by the philosophy of his history, by a search for ever-greater reality and truth. When he succeeds, he does so by following tradition, and when he fails, he fails because he has strayed from its teachings.

Beyond the instant, beyond the historical man and the historical age he lives in, his consciousness, it would seem, is suffused in the truth of that reality that is found at the core of the lives of our heroes of history.

Man is not insulated from history; he is constantly made aware of it through the lives of those who lived before us and who continue to live in us, filling time and space with examples of their deeds, which we may meditate upon.

History is the adventure of mankind, shaped by man's thought, which is itself determined by the particular community or people concerned.

The animist ponders whether history is a sequence of isolated events or the true life of man when he is integrated into himself and with the universe.

History is in truth the existence of this universe and its evolution, which is governed by patterns repeated by the prime movers, whose lives govern our own lives.

Is time really the instant which we experience, or is it rather a continuous existence which, beyond the dimensions of space and time, is experienced differently in different contexts? Do these contexts suggest changes and mutations which may have modified the real state, time and space at the level of the intermediary world, the true world to which the animist's view of life is keyed?

The animist's philosophy is fed by sources that lie equally deep; and these sources give a different meaning to history.

Over and above the teaching which it is their function to give, the priest, the initiator and the sage are exemplars for living to which succeeding generations may look up.

What is man, what is a race of men but a succession of moments, whose history, whose failures and successes are embodied in the lives of our outstanding historical figures?

Truth. History is absolute time dissolved in the infinity of space. It is the historical contingency, in which man rises up from temporal time and makes his presence known, a presence bound up with the life of his people.

It is this survival of the species in each one of us which governs the course of history.

History, being made by and for man, must serve man's cause, for it cannot be distinguished from him. It explains and justifies his actions in terms of the aim to be achieved in the accepted order, consonant with the harmony of the universe which sustains him.

To the animist, history is above all a phenomenon which implies man's constant presence in the historical process, shaping it to some moral purpose, gearing it to the service of the individual or the community at large, in brief, to the service of mankind.

Despite the diversity of our outstanding historical figures, they all draw their inspiration from the inviolate substratum of Africa. Such ancestors as Sundiata Keita, the Mansa Kanku Mussa and Suleiman, Sonni Ali Ber, Askia, Samory Touré and Behanzin are not only of their own time, they come vividly to life and bring their times vividly to life in the traditional tales told by the *griots*. They remain the soul of their peoples, which draw a new lease on life from the memory of such figures, finding in it their reason for living and the courage to mould history to the shape of their era.

The wisdom of the ancients continues to govern the present and future. It is there, intangible, through time and space, in the same way as our soul, faced with the contingent circumstances and difficulties of our individual, daily lot, remains unchanged.

The astronaut who watches the sun rise twenty-four times in a single day is living in the world of reality, for it is his achievement to have circled the earth at such a speed that he can live twenty-four dawns in one day.

Were our human body to live through this same adventure we should succeed in transcending both time and space and in seeing things that no man has yet seen in a relative context.

The animist believes that, however paradoxical it may be, this question is within the domain of N'Debi, within God's power to reveal all things to His elect.

It is this fundamental reality that the priest, the soothsayer and the prophet claim to reify, these three to whom God, in His greatness, reveals the secret which hides the future from us. Thanks to their creative imagination they are able to enter a concrete spiritual world and to trace its mutations which, since they are not scientific in nature, are not governed by the laws of science.

The animist nevertheless believes that man finds nothing that does not already dwell in him. Whereas what we call his magic conjures up the object, science, which is external to man, discovers and reifies it in the absolute duality of this man and his creation—which sometimes indeed rises up against him.

This may indeed be what is troubling the present-day consumer society, in which man has lost all sense of his own identity.

This is why it is wrong to destroy a civilization, which is essentially man's work; it can help us to correct our course, when we stray in our development from the truths that govern our humanity. Although himself no fatalist, the animist appears to believe in destiny. He sees it in the past and the future, and experiences it in the present, knowing that he is continuing the same process in different contexts.

Man's role in the physical and spiritual universe is to make his life harmonize with the necessary balance of nature—many of whose forces still remain a mystery to us.

This conception of man and life justifies the animist practice of initiation, in which the initiate is revealed to himself, his good or ill-fate is made known to him, and that wisdom is instilled in him that will enable him to enjoy his fate with moderation or to bear it patiently, in the knowledge that he does so as part of the common lot of his clan or his people.

The animist in his profound wisdom situates the whole being in the world's destiny, which subsumes that of the human species, the sum of the continuous efforts of the generations that succeed one another in unending time—which we call past, present and future.

The truth is that we have been, are and shall in all circumstances continue to be what our joint efforts have made of us.

For the animist, history is that philosophy of life which postulates a union of being and space.

Its security-bringing process determines our myths outside the realm of science fiction and the fantastic. It occupies the sphere of man's consciousness. The purpose of the historical process is to show him, whenever he goes astray, a new and sure path to take.

NOTES

1. The seven upper heavens.
2. The seven lower heavens.
3. According to my master Amadou Hampâté Ba, this demiurge is called Endebi among the primitive Fulani, a name very close to N'Debi.
4. In animist doctrine, a tree or man can be destroyed as follows: (a) a tree—by destroying a leaf or some other part, the tree's double is destroyed, and this results in the death of the tree itself; (b) a man—through a hair or some other part of the man's body, it is possible to reach the man's double, whose death brings about that of the man himself. This means that the physical is part of the spiritual whole and vice versa.
5. This distance implies the space and time needed to cover it.
6. When using magical formulae, the priest avoids pronouncing the name of the lion, the snake or any other dangerous animal from which he is trying to save the person who has asked his help or for whom he is interceding. Here, the lion (*musu*) is called Kotombi, literally 'he-who-grapples'.
7. *Dagar* is an onomatopoeic word indicating the restless mood of the *Sa* (the angry lion).

8. *X* being the object of the lion's wrath.
9. The genii are doubles but, like men, they are mortal. At their death they vanish into the ground in the same way as spilt water.
10. The term 'double', meaning another dimension of man, is a better term to use for this dimension, which appears to be a projection of the integral being on a more evanescent plane of being.
11. What I am suggesting here cannot be understood by focusing the imagination of the visible, concrete world, but only by considering it in the light of the creative imagination, in which the spirit seeks to apprehend a perfectly 'objective' existence.
12. The iron dross and the fragment of old *canari* symbolize the time which it is desired that one's undertaking shall last.
13. These *Bisimilahi* are not followed by *Rahamani rahim*. They are focused solely on this world, to which the animist confines his material and spiritual action. They doubtless derive from pre-Islamic Arabia or Yemen.
14. 'It' being the secret, the formula.
15. The genii or gods, the shades of the ancestors.
16. Men, soothsayers, seers and mediums.
17. The major obstacle.
18. I have put it out, that is, I have exorcized the evil.
19. At present, only my hand, that of N'Debi is upon you, and it is upon you for ever.
20. Such figures are encountered frequently among the Sufi, the freemasons and in esoteric doctrines.
21. Inhabitants of the banks of the River Niger, fishermen, who are possessed with special powers related to water—'masters of the waters'. The name changes from region to region.
22. *Textes Sacrés de l'Afrique*, p. 8, 9 (Foreword), Paris, Gallimard, 1965. (Coll. Afrique.)

THE PROPHET

Louis Gardet

The mere notion of prophecy implies a very special relationship with time, a break in the inevitable time sequence, an irruption into daily life of the points where time and a world beyond time meet, a reading of the facts of history in a light which transcends them.

Prophetism lies at the heart of both Judaeo-Christianity and of Islam; the lines cross but retain the specificity they derive from their different mental horizons. We are not concerned here with the whole theory of prophecy, its necessity or use, the role of the prophet, his mission, the nature of revelation. Prophecy developed in Judaism, Christianity and Islam, each in its separate way, and is one of the great theological themes in all three. Our more modest purpose is to consider the time experience once prophecy is accepted by regarding the prophet as seen by each of the great monotheistic religions.

The prophet and the presence of God in human time

First, the points which the three religions hold in common.

The same word *nabī* is used to designate a prophet in Hebrew and Arabic. In Hebrew, the *nābī'* is he who tells men what God bids him to tell them, and God 'is with the mouth of the prophet'.[1] In Arabic, the root *nb'* has come to have the general sense of arriving or appearing, and, secondly, of announcing news. In the secondary sense, the *nabī* usually means he who announces something to men on behalf of God, and consequently he whom God entrusts with a mission to men, 'he whom God-Truth [*al-ḥaqq*] sends to his creature [*al-khalq*]'.[2]

In the European languages, prophet (in Greek, *prophētēs*) has the stricter meaning of 'he who tells the future', who reads in the future events yet unknown; in Christian theology, accordingly, a 'prophetic light' may be transmitted and received, if God so wills, by someone who is not at all God's witness, e.g. Balaam[3] and Caiaphas.[4] In such cases the prophetic light is transitory: there is no real prophetic mission, which is not merely a prediction of events to come (even if profoundly understood, e.g. the words of Caiaphas interpreted by St John),[5] but part of God's design which makes a man the witness of his presence and confers on him something of his mystery and his doing in the created world.[6] It implies divine election, a man freely chosen by God to be his envoy. Contrary to a whole trend of philosophical interpretation, to which we shall return, man becomes a prophet not by natural qualities but by the free choice of God.

Here, the Bible, Gospels and Koranic faith all agree. 'And I have chosen thee for Myself' says God to Moses in the Koran (20: 41). The very fact of this choice implies that prophetic enlightenment must be combined with what may be termed prophetic grace: to be God's envoy the prophet must be wholly true and sincere, unfailing, vigilant in his mission. He may have lapses, but no account exists of a prophet authentically sent by God who betrayed his trust. The 'false prophets' are false from the outset either because they trust in human subterfuge or succumb to errors induced by following 'false gods'. God's hand is on his prophet, whom he protects from error.

Traditional Muslim thought exalts this divine protection into a privilege of sinlessness (*'iṣma*), for the duration of the mission according to the 'best' authorities, from birth to death according to others.[7] According to Islam, the prophet thus escapes the major failings of humankind and this, in the eyes of the doctors, justifies popular veneration for the prophets, and above all, for Muhammad.

God, through his messengers, intervenes visibly in human history. If a prophet can predict, that is proof that the ineluctable irreversibility of past, present and future is breached, and attests the relativity of this endless rhythm, the irruption into the daily round of moments of eternity that are incommensurable with time. The prophet lives both in numbered, successive time and in a spiritual time which is the projection of divine eternity in its 'unenduring duration'. For him it may be said that the two types of duration coalesce, hence the symbolic expressions punctuating the announcement of future events foretold or threatened. The 'weeks' of Daniel (9: 20–27), the 'thousand years' of the Apocalypse (Rev. 20: 1–6), the imminence of the Day of the Lord (Joel 1: 15; 2: 1–2) (cf. the Last Hour proclaimed in the Koran by the Meccan *suras*) refer to a qualitative time, more real than the regular sequences which reason measures, but which continuously upsets reason and forces it into arbitrarily precise interpretations.

The prophet becomes a prophet in continual dependence on the divine initiative. This distinguishes him, absolutely, from the magician or soothsayer, who also aim at transcending time, but on their own initiative and by the mastery they seek to acquire over the laws of the cosmos. The initial reaction of the prophet, on the contrary, is nearly always a lively fear, and then humility, when faced with the divine *tremendum*, the overpowering numinous world. One need only read the mission of Moses[8] in the Bible, the call to Samuel or Isaiah, the protest of Jeremiah.[9] The Word of God which 'falls' on the prophet is 'weighty' (*thaqīl*) according to the traditional Muslim commentary on verse 73: 5 of the Koran. The Koranic injunction 'Say: I am only a mortal like you' (18: 110) rings out like an echo of Jeremiah (1: 6): 'Ah, Lord God, behold, I cannot speak: for I am a child.'

For the prophet is the servant of the Word. God sets him apart, consecrates him and so empowers him to communicate God's message to men. The mother of Samuel has 'lent him to the Lord as long as he liveth',[10] and God solemnly states to Jeremiah (1: 5): 'Before I formed thee in the belly I knew thee; and before thou camest forth out of the womb I sanctified thee, and I ordained thee a prophet unto the nations.' Similarly, Islam lays stress on the 'sending' (*risāla*) of the prophet. The greatest of the prophets, those whom God destines to be the law-givers of their people, are his apostles, messengers (*rusul*). To be 'the Messenger of God' (*rasūl Allāh*) is Muhammad's great title of honour.

The Word of God, because it creates *ex nihilo*, is beyond any time sequence. It *makes* time and dominates it. 'God said ...' (Gen. 1); 'He says to it only: Be [*kun*], and it is' (Koran, 2: 117). It was by his Word that God created all living things, at that moment of eternity which is absolute beginning. Consequently, whenever God, through his prophets, addresses his Word to men, earth and heavens open on this primordial creative moment. Like a sharp-edged sword the Word of God transpierces time's break, the immeasurable gulf of radical otherness between Creator and creature. Before the gaping abyss the prophet is terrified, vulnerable, for often he is just a poor and simple man; but he is as though hurled by divine election and consecration into this gulf; he judges the numbered time of men and measures its successive duration against God's absolute 'Be'.

And so prophets, at least at first, are not welcomed by their contemporaries, whose habits they upset. The first reaction of men, hidebound in the everyday, eating, drinking, sharing earthly hopes and pleasures, is to ignore the hard summons from God. Each prophet of biblical times was tested. He is criticized and persecuted: 'A fowler's snare is on all his ways, and hatred in the house of his God' (Hos. 9: 8); and the Gospels: 'For so persecuted they the prophets which were before you' (Matt. 5: 12); and still more vehemently: 'Therefore also said the wisdom of God, I will send them prophets and apostles, and some of them they shall slay and persecute:

that the blood of all the prophets, which was shed from the foundation of the world may be required of this generation' (Luke 11: 49–50).

It is in the Gospels that the tragic destiny of the prophets is most forcefully emphasized. The theme is taken up again in Islam. The opposition of the Meccans to Muhammad leads to his exile (*heira*) to Medina. 'The prophet is a fool, the spiritual man is mad' said the adversaries of biblical times (Hos. 9: 7); and the unbelieving Meccans: 'Medleys of dreams: nay, he has forged it! nay, he is a poet' (Koran, 21: 5). It is the lot of all God's Messengers to be treated as impostors.[11] All prophets—the theologians of Islam will teach—have first to face the incredulity of their fellow citizens, which they 'challenge' (*taḥaddī*) with the 'prophetic miracle' (*mu'jiza*) to convince them of their error. Thus can an authentic prophet be recognized. (Muhammad's exemplary 'prophetic miracle' was the Koran itself.) We shall later return to certain differences in outlook, in Judaeo-Christianity and Islam, regarding the fate of prophets.

The opposition to God's Messengers is not so much against God as against the intrusion into history of his eternity, which is incommensurable with time. For the prophets, witnesses of the presence of God, shatter as it were the tacit and reassuring assumption of second causes.

The prophet and the future: Judaeo-Christian trends

The preceding remarks apply by and large to the prophet's attitude to time in Judaism, Christianity and Islam, but each of them will regard it in its own way.

One major difference is that in the Jewish and Christian faiths the prophetic message is pregnant with a future which human reason could not have foreseen and which transcends all earthly hope. There is a divine *oeconomia*, a divine plan for the race of men. The Word which God sends them secretes a gradual revelation. The relations of God and men are inscribed in a Covenant or Covenants (in Hebrew, *berit*) which God offers, which his people freely accept and which delineate the broad outline of a history.[12] And God who is 'gracious and full of compassion [...] will ever be mindful of his covenant'. He 'hath commanded his covenant for ever' (Ps. 111).

His messengers are the keepers of the covenant; they repeat its demands and its benefits, and gradually reveal its hidden content. They pronounce God's judgement on the incredulity and faithlessness of men but they carry a message of hope—the hope of salvation and the return to grace. For the first task of the biblical prophets is to announce the Messiah and Saviour. Their predictions which are closer in time—victories and defeats, the return of the scattered peoples—are merely the precursory signs of the Messiah's coming, its image as it were. The biblical prophets are the heralds of the coming Messiah. The various sequences of events are merely landmarks along the road towards that trans-historical moment when the promise will be fulfilled. Even the choosing of Israel, 'Israel whom I have chosen' (Isa. 44: 1), is the annunciatory image of the Messiah, the universal Saviour, to whom Israel is to give birth.

Appeals for spiritual conversion and the interiorization of the working of God become increasingly pressing as biblical times advance. For not having received the Word of God in its inmost heart Israel will be punished. Anathemas are pronounced by the great prophets against 'a stiff-necked people'. But in his heart of hearts the prophet shares the misery of his people (Jeremiah) and constantly intercedes for them (Moses, Isaiah, Jeremiah, etc.). Foreign 'nations' are God's instrument for punishing his people; woe unto them if they triumph, however, for God's wrath will ultimately descend on them (Deutero-Isaiah, Jeremiah, Ezekiel). The prophets prophesy 'against' the nations. They do not prophesy against Israel, but announce the punishment that is to come if it does not repent, or proclaim the real meaning of a punishment already suffered, which is a call to repentence and hope. 'O Israel, return unto the Lord thy God for thou hast fallen by thine iniquity' (Hos. 14: 1).

The inner conversion demanded by the prophets applies as much to relations between men as to man's relations with God: hence a constant defence of the poor and the oppressed, a constant demand for social justice. The pure worship of God and the love of all justice are one and the same thing. 'Rend your heart and not your garments, and turn unto the Lord your God, for he is gracious and merciful' (Joel 2: 13)[13] and 'I hate, I despise your feast days ... your offerings, I will not accept them ... but let judgement run down as waters, and righteousness as a mighty stream' (Amos 5: 21–4). And with even greater insistence: 'Is not this the fast that I have chosen to loose the bands of wickedness, to undo the heavy burdens, and to let the oppressed go free, and that ye break every yoke? Is it not to deal thy bread to the hungry, and that thou bring the poor that are cast out to thy house; when thou seest the naked that thou cover him ... ?' (Isa. 58: 6–7).[14] The prophet announces the future but also guides his people in their daily life. He bears witness to a future which is already present, which must gradually enter the personal and social life of men and against which political leaders and false prophets constantly baulk and protest.

Thus the biblical prophet at one and the same time partakes of the transcendent duration of God and is immersed in time and its successive manifestations which between them weave the fabric of human lives. For this temporal succession, in its innermost sense, already connotes an eschatological dimension. It is not only the announcement of the coming of the Messiah; it already takes on Messianic value, to the extent that the 'circumcision of hearts'[15] makes it possible to receive God's call. The prophet is, indeed, the herald of the Messianic future, but by the same token he is also the sentinel ('Watchman, what of the night?') (Isa. 21: 11) who keeps watch upon the towers of the city to warn the people.[16] He is both outside and inside time. The prophetic judgement on human existence no doubt involves eschatology but it also directly involves history.

A difference enters at this point between Christianity and post-biblical Judaism. According to the Christian faith, the Messiah came in the person of Jesus of Nazareth. The Evangelists are constantly concerned with proving that in him the preceding prophecies are fulfilled. Human time, to which the announcement, threats and promises of God's messengers alone gave authentic meaning is, as it were, recapitulated in and by the Gospel of Christ. The promise is fulfilled and now 'the kingdom of God is within you' (Luke 17: 21). However long it may be, the ensuing succession of centuries[17] is the final time of the world, preceding the Apocalypse of the last days and the Second Coming, this time in glory, of Christ the Redeemer.

Christ, dead and resurrected 'for our sins and our salvation' is, according to the Christian faith, more than a prophet; he is the Word Incarnate. But he is also, superabundantly, a prophet.[18] He prophesied the destruction of Jerusalem, the end of the world and his Second Coming. He belongs to God's eternal and uncreated duration; moreover, he receives prophetic illuminations in his human soul, intimations arising from divine eternity which make God's messengers the bearers of his Word. The biblical prophets guided the people towards the Messianic age. This mission comes to an end with the coming of Christ who is the Eternal Word and in whom all is fulfilled. 'Think not that I am come to destroy the law, or the prophets: I am not come to destroy, but to fulfil' (Matt. 5: 17). The spirit of prophecy will continue to imbue and dwell in the Church of Christ, and in all the redeemed who remain faithful to God; the times of the prophets are fulfilled.

Judaism did not subsequently recognize Jesus as the Messiah. For it, therefore, the times of the prophets remain, should remain open (as certain trends of Jewish mysticism would readily agree). But the destruction of the Temple and the Diaspora seem to have created a new status, requiring Israel to live in absolute fidelity as the chosen people to the commandments of the biblical age while awaiting the Messiah to come. 'The end of prophecy does not announce the end of history.'[19]

After the destruction of the Temple Judaism entered the age of glosses and commentaries. Possible new revelations are not at all the aim of the ceaseless work of its scholars on the biblical text. For Israel too, but for reasons very different from those of Christianity, the times of the prophets are over. Modern thinkers, however, such as Edmond Fleg, have attempted to make of the Jewish people itself, in its historical existence, the awaited

Messiah: its destiny among the nations, its unbelievable sufferings, and its never abandoned hopes assume prophetic value. There is a strong temptation here to render earthly history itself sacred. It may be added that many representations of Jewish spiritual thought disagree and continue to live outside eventful time, as if time were suspended, and the 'return' prophetically announced in former times could be acknowledged only if it took a directly Messianic form.

Islam: the prophet and discontinuous time

The Moslem faith has its origins in the factual existence of prophecy, and, at its most strict, affirms the Koran non-created, being the actual Word of God, as God communicated it, word by word, to the Prophet. The Koran constantly refers to previous prophets, principally those of biblical times,[20] and certain others, too, whom God sent to peoples—'Ād and Thamūd—who did not receive them. From this point of view Islam has a 'sacred history' in which God intervenes in human time by occasionally sending messengers with warnings. But all previous revelations are recapitulated and elucidated by the Koran, for Muhammad is the 'Seal of the prophets' (*khatm al-anbiyā'*) (Koran, 33: 40), responsible for transmitting the ultimate teaching and the ultimate religious law sent to men by God.

Left to itself, mankind becomes careless, forgetful and ignorant (*jāhiliyya*) of God, plunged in darkness that is sent like lightning flashes by the missions of the prophets and envoys. It was to triumph over the night of the *fatra*(*s*) (intervals) that God, in his mercy, made Muhammad his final envoy, charging him to establish the 'best nation raised up for men' (Koran, 3: 109) which faithful to his Word will, with divine help, endure until the Last Hour. This Word, searingly epitomized in the *shahāda*, the testimony of Islam, is in essence the affirmation that there is but one God, One and transcendent, and that Muhammad is supremely His Prophet.

The Koran, of course, rescinds the 'law of the Gospels', which itself rescinded the Tora. It rescinds them but at the same time and by the same token restores to them their original meaning which had been obscured by the infidelities of the Jews and the Christians. For, while Islam has a 'sacred history' there is no true divine *oeconomia* or gradual revelation. In its inviolable essence, Koranic teaching is the reaffirmation, pure and simple, of the absolute and unique Lordship of God. Thus it repeats, for this day and age, the message which each of the prophets had for his own contemporaries; it re-experiences, in its full intensity, the faith of Abraham 'the friend of God'. It introduces into human time the great meta-historical proclamation of the 'pact' (*mīthāq*) of pre-eternity.

This is no longer the free and mutual acceptance of a Covenant, but a pact which the Most High grants to his creature and sets like a seal, a sign of belonging, on the hearts of all men. When God, in the pre-eternity before the body was created, 'brought forth from the children of Adam, from their loins, their descendants, and made them bear witness about themselves: Am I not your Lord? They said: Yes; we bear witness. Lest you should say on the day of Resurrection: We were unaware of this' (Koran, 7: 172).

The ages which separate the Creation of the World from the Last Hour bear, of course, their burden of events but find their true significance only by referring back in this way to the primordial pact. The Koranic prescriptions represent less an advance in divine revelation than an elucidation of the past, a guarantee of its permanence, provided the 'Community of the Prophet' remains faithful to them.

This everyday presence of faith and testimony no longer leaves any room for biblical expectations, hopes and promises. In a sense, any prophetic revelation or any word transmitted to men by God brings time to a stop. It might be said that in Judaeo-Christian thought the prophet stops time in anticipation of a future already present and still to come; in Islam, the prophet stops time by an immobile return to the primordial moment which is that of the pre-eternal pact, of Abraham destroying the idols, of Muhammad exiling himself to Medina, even that of the Last Judgement. The biblical prophet, God's witness, is also a witness to progress in human history; the prophet of Islam appears as a witness to the unique moment (*waqt*) of eternity, in which the destiny of man becomes as naught in the presence of God.

This absence of a historical dimension in the ordinary sense of the term may perhaps explain the different ideas of revelation held by Christianity and Islam. According to Christian theology, divine revelation, the Holy Spirit, is expressed in and by the freedom of the prophet, by his own words and his individual style. In Moslem theology, revelation is the Word of God 'supernatural dictation',[21] which the prophet merely repeats faithfully, word for word. The prophet is not a free instrument, and no exegetic question of historical descent or source can arise.

The biblical prophet is the guide and conscience of his people but, since Saul and the introduction of monarchy, not necessarily its leader. A distinction already drawn in the bible between spiritual and temporal will be underlined in the Gospels. For human time, stopped through and by the very existence of the prophetic mission, nevertheless continues to flow past on its own second-cause level and retains its power to create effects. In Islam, the prophet is by right the head of his community as long as he lives. As we have seen, the Koran also stresses the opposition and incomprehension he always meets at first. But unlike the Gospels it does not say that prophets will be 'slain and persecuted'. This opposition to God's envoy can and must be a passing phenomenon only. Muhammad had to go into exile from Mecca, but he returned a conqueror and a conqueror by force of arms. The biblical prophet is not a war leader; the prophet of Islam was. We shall not revert to the perception of discontinuous time, the 'constellation, a Milky Way of instants'[22] of the Moslem theologians. Suffice it to say that prophetic time and discontinuous time here coincide and that any event may acquire the value of a portent. The time of human history and its tangled skein of second causes loses its particularity to become, in its discrete moments in time, a pure manifestation of the divine commandment (*amr*). Hence, apparently, the fusion of spiritual and temporal peculiar to Islam.

Other differences may emerge during this post-prophetic 'waiting period'[23] which the community must live through while awaiting the Last Hour. According to the majority Sunni tradition, nothing was actually arranged during the Prophet's lifetime, or by the Prophet himself about the succession at the head of the State of Medina; this is possibly one of the main points of difference with the Shī'ite opposition who affirm, on the contrary, that texts exist which settled the succession in favour of the *Imām*, who must belong to Muhammad's family. In fact, the Shī'ite *Imām* enjoys prophetic privileges: prophet-like he transcends time, and through his knowledge of 'hidden meaning' and his total freedom from sin he participates in divine wisdom and duration. And special prestige comes to attach to the twelfth *Imām* of Twelve Shī'ism,[24] 'vanished' (*ghā'ib*) yet still living and who will reappear when the world comes to an end.

Sunni Islam does not accord the *Imām* any such privileges. But for it, too, prophetic time is the underlying reality of numbered and successive time in its discontinuous sequence. Prophetic time continues to impinge, tangentially, on the life of the community. It constantly recalls that the Last Hour is at hand, even if no man can tell when it comes, in the never-abandoned anticipation of the *Mahdī*, the Well-Guided, who will come in the name of the Prophet to guide the community on the 'straight way'.

The worlds beyond time

The intervention of prophets in history, whether in Jewish, Christian or Moslem thought, thus takes the form of a sundering of the categories of time, an easing of the weight of days in a moment of transcendent duration. The initiative does not come from men, it does not come from the prophet, it comes from God. Prophetism finds its real place only in a monotheistic and creationist vision of the world, one in which a Living God transmits to created man the Word of salvation.

The Arabs before Islam dreaded the weight of the continuous duration of days, of the overwhelming duration (*dahr*).[25] There is no need to stress how deeply the anguish of time, the 'watchful and deadly foe',[26] as Baudelaire called it, is engraved in the hearts of men. It forms a poignant strain in modern poetic experience: *Mutter Erde! rief ich, Du bist zur Witwe geworden, dürftig und kinderlos lebst Du in langsamer Zeit* (O Mother Earth, I cried, Widowed thou art, needy and childless in the slow march of time (Hölderlin));[27] while contemporary thinkers and philosophers such as Proust, Bergson or Heidegger are haunted by it. This anguish, in its acute forms, is perhaps the product of a culture in which man is left face to face with himself and himself alone, abandoned to the mortal weight of *dahr*, ignorant of or rejecting the ancient visitation of the prophets ...

It is noteworthy, moreover, that religious beliefs which are not theistic in themselves, or at any rate not creationist, have had no true prophets but have nevertheless sought in a world beyond time their particular spiritual experience. Here it is not so much a sundering of human time that is involved as the transition to a different mode of duration, e.g. Hinduism and Buddhism. Both have a Holy Scripture, in Hinduism a timeless revelation,[28] in Buddhism the fruit of the experience of the Enlightened One. The Enlightened One, the Buddha, is not a 'prophet'; he is not the Apostle of a transcendent God. It is at the peak of his own inner ascension, and through it, that he achieves Enlightenment. The inspired sages of India perceived timelessly the sacred text of the Vedas not as the Word of a Living, Creator God but as the expression of a truth which is cosmic and trans-cosmic at the same time. And it was the transparent character of a Self all of whose separate determinations had been sloughed off which enabled them to partake of this 'hearing' (*śruti*).

It may be said that the experience of yoga, or its Buddhist equivalents, is a search for a world beyond the contingent web of space and time,[29] a laying oneself open to the spiritual (meta-spiritual) forces of man, albeit counter to the ordinary mode of knowing. Does not this hope for a world beyond time (and space) lie at the very heart of the popularity which yoga and Zen currently enjoy in the West? And is this detour via the great cultural heritage of the Orient the only means whereby the monotheistic traditions are to rediscover that searing of time by divine eternity which was once provided by the voice of their prophets?

The Hindu and Buddhist traditions may offer the best reply that man is capable of giving to himself. The prophet sent by God brings God's reply. In both cases we are confronted with a world beyond successive time, whether continuous or discontinuous. But in Hinduism and Buddhism we are concerned with an immanent way by which man transcends himself through the sole light of the absolute which is within him; and numbered time is abolished in a spiritual and incommensurable duration which is not yet eternity. In the prophetic revelation all initiative comes from God; divine eternity pierces through human time to reveal, in the blinding light attending the visitation, its authentic significance.

In closing we shall note that the monotheistic traditions at one stage produced an explanation of prophetism which in a sense would seem to tie up with the great experience of the Orient. It was in the Moslem cultural zone that this developed, under directly Plotinian influences.[30] We here refer to the theory of prophetism developed by the *Falāsifa*, the Hellenistic philosophers of Islam, particularly the oriental *Falāsifa*, Fārābī and Ibn Sīnā (Avicenna).

In them the way of immanence and self-transcendence of the *Enneads* merged with the Koranic teaching on the Prophet and the prophets. The point of fusion came through and in their theory of knowledge. For them, the prophet is a prophet *by nature*. He is a man whose intelligence is so lucid, whose power of imagination and mastery over matter are so strong, that in this life he can attain the world of separate Intelligences and pure Intelligibles. From it he will receive the higher light, itself an emanation from the First Being, who is God.[31] Making the necessary monotheistic transpositions, we are quite close here to the 'hearing' of the great sages of India.

We should add that for the *Falāsifa* all this is seen from the standpoint of an emanatist view of the world—an eternal emanation which comes necessarily from the First Being and is willed by him.[32] Thus the whole sphere of existence is rigidly determined and there can be no place for a free contingent future.[33] Consequently, the prophet is no longer 'a mortal like you', freely chosen by God and aware of his grace. Because of what he is by nature, he can only be a prophet; by nature he has the power to give men the teaching which they need. Time is no longer sundered or broken apart because time, which can only be understood in terms of analogy, is eternal. As in Hinduism and Buddhism, there is passage from one type of duration, that of corporeal beings, to another, the incommensurable duration, of the spirit.

This view of the *Falāsifa* did have some sporadic influence on the Muslim concept of prophetism,[34] although the strict 'reformers' protested against any theory which, in the guise of honouring the prophet, raised him to a superhuman level. The attempt of Ibn Sīnā to incorporate into his system the Muslim concept of prophetism was well known to Maimonides, who did not accept it. Thomas Aquinas placed it in a line which he termed that of 'natural prophecy', and which he clearly distinguished from authentic prophecy imparted by God. An echo of the views of Fārābī and Avicenna can be found in Spinoza (perhaps through the intermediary of Maimonides?)

We are far removed here from the prophets of biblical times or the text of the Koran. To recapitulate, in the line of Koranic teaching as developed by Muslim thought, the prophet is the sign of a break or shear in the successive and discontinuous duration of numbered time, brought about by the creative Word of God. The biblical prophet—and Christ in his role as prophet—is both at the very heart of human time and on the line of shear, at the points where time and eternity meet. The prophet is the sign of a transcendent Presence which delivers human duration of its burden of mortality and causes it to participate, not by nature but by grace, in the eternal youth of the 'unenduring duration' which is that of God.

NOTES

1. cf. Exod. 4: 11–12. (*Note:* All biblical quotations are taken from the Authorized (King James) or Revised Standard Versions).
2. Taftāzānī, *Maqāṣid,* p. 128, Istanbul.
3. cf. Num. 22–4.
4. John 11: 49–52.
5. 'You do not understand that it is expedient for you that one man should die for the people, and that the whole nation should not perish' says Caiaphas about Jesus (John 11: 50).
6. 'What he has to say is not a prediction; it is given immediately at the moment of speaking. In this prophesying, vision and speech alike seek to lay bare. But what they reveal is not the future but the absolute. Prophecy answers to a longing for knowledge, but not the knowledge of tomorrow, the knowledge of God ... It implies some sort of relationship between eternity and time' (André Neher, *L'Essence du Prophétisme*, p. 9, Paris, Calmann-Levy, 1972).
7. See, for example, Louis Gardet, *Les Grands Problèmes de la Théologie Musulmane: Dieu et la Destinée de l'Homme*, p. 184–8, Paris, Vrin, 1967.
8. Exod. 3: 9–12; 4: 10–7.
9. cf. 1 Sam. 3: 1–18; Isa. 6: 5; 29: 11; Jer. 1: 5–8, etc.
10. 1 Sam. 1: 28.
11. Koran, 6: 34; 35: 4; 3: 184. *Note:* all references to the Koran are taken from the English version by Maulana Muhammad Ali.
12. 'The *berit* is a divine plan in the two senses, psychological and geometrical, of the term. In offering it, God is pursuing a precise intention (the binding character of the Law) and enlisting the help of men for its ultimate achievement (the participatory character of the Covenant). (Neher, op. cit., p. 144.)
13. cf. Jer. 7: 21–3.
14. cf. Isa. 61: 1–3.
15. This theme recurs very frequently in the biblical prophets and is taken up by St Paul.
16. cf. Ezek. 33: 2–6.
17. Which is the 'time of the Church'.
18. cf. Thomas Aquinas, *Summa Theologica*, IIIa, q.7, a.8.
19. Neher, op. cit., p. 234.
20. These prophets are not, in fact, identical with those of the Bible. The Koran names very few 'prophets' in the true sense, rather great patriarchs and kings.
21. Louis Massignon, 'Situation de l'Islam', *Opera Minora*, Vol. I, p. 16, Beirut, Dar al-Maaref, 1963.
22. Louis Massignon, 'Le Temps dans la Pensée Islamique', *Opera Minora*, Vol. II, p. 606. For this apperception of 'discontinuous time', see our study: 'Moslem Views of Time and History—An Essay in Cultural Typology', *Cultures and Time*, p. 197–227. Paris, The Unesco Press, 1976.

23. 'It is through this idea of a "waiting period" that the notion of "duration" entered obliquely into Islamic thought.' (Massignon, 'Le Temps dans la Pensée Islamique', op. cit., p. 607).
24. Which, since the sixteenth century, has been the official religion of Iran.
25. Koran, 45: 26.
26. Beaudelaire 'Le Voyage', VII, *Les Fleurs du Mal,* p. 159, Paris, Garnier Frères, 1961. cf. 'L'Ennemi', ibid., p. 18.
27. 'Der Wanderer', *Elegien.*
28. 'These texts, transmitted orally with amazing accuracy, are regarded by orthodox Brahmanism as eternal and uncreated, literally "without beginning" and "non-human". They were perceived and communicated in human time by inspired sages; as they are essentially verbal, they must be "heard" to be known: consequently they are referred to collectively as the "hearing"' (Olivier Lacombe, 'Le Brahmanisme', in the collective work, *La Mystique et les Mystiques*, p. 731–2, Paris, Desclée de Brouwer, 1965.)
29. 'Yoga is a practical technique which claims to change consciousness in such a way that the yogi can experience a state of being which transcends space and time.' R. C. Zaehner, *At Sundry Times* (re-translation of the original English from French).
30. Influences exerted via two texts translated into Arabic and wrongly attributed to the Stagirite: (a) The 'Theology of Aristotle' (*Kitāb uthūlūjiyā Arisṭūṭālīs*), being extracts from the *Enneads* IV–VI and a fragment of 'Porphyry' somewhat glossed at times in the translation; (b) 'Aristotle's Exposition on the Subject of Pure Good' (*Kitāb al-īḍāḥ li-Arisṭūṭālīs fī l-khayr al maḥḍ*) known in medieval Latin as *Liber de Causis*, being extracts from the *Elementatio Theologica* of Proclus.
31. For Fārābī, see, for example: Richard Walzer, 'Al-Fārābī's Theory of Prophecy and Divination', *Greek into Arabic*, p. 206–19, and refs., Oxford, 1962. For Ibn Sīnā, cf. F. Rahmān (ed.), *Shīfā', Kitāb al-nafs*, p. 160 et seq., Oxford, 1959; and *Ilāhiyyāt*, Vol. II, p. 435 et seq., Éditions du Caire, 1380 H. (1960). A parallel text in *Najāt*, 2nd ed., p. 292 et seq., Cairo, 1357 H. (1938). See also *Ishārāt*, Forget edition, p. 209–20, Leyden, Brill, 1892.
32. Mode of 'willed necessity', in conformity with the Plotinian adage *ex uno non fit nisi unum,* but intended to safeguard the exclusion of extrinsic motivation in the production of beings; cf., for example: Ibn Sīnā, *Sharḥ Kitāb Uthūlūjiyā* [Glosses on the Pseudo-*Theology of Aristotle*], in Badawī (ed.), *Arisṭū 'ind al-'Arab*, p. 63–4, Cairo, 1947.
33. Hence the famous text: 'If a man could know all events and phenomena, of earth and sky, and their nature, he would know for certain how will come about all things that will come about in the future' (Ibn Sīnā, *Najāt*, p. 302).
34. A tendency emerged—in Ibn Khaldūn himself—to see in prophets superior men, placed by nature at the peak of (or above) the human race.

THE GURU AS EXEMPLAR OF AND GUIDE TO THE TERM OF HUMAN EVOLUTION

Sri Madhava Ashish

The primary meaning of the Sanskrit word *gurū* is 'heavy' or 'weighty'. By derivation it comes to mean 'great', 'respected', 'venerable'.

In the pantheistic culture of Hinduism all things are divine and all acts sacramental. And since such a culture does not draw hard and fast distinctions between objects and the divine content of objects, he who possesses the sacred knowledge, teaches it, teaches the sacred language in which it is enshrined, teaches the sacred texts, and initiates into the sacred rites, is the guru, great, respected, venerated, and worshipped as identical with the knowledge he expounds. Similarly, no absolute distinction is made between mundane knowledge and transcendental knowledge, because the divine is both immanent and transcendent. The teacher of a craft is thus as much guru as the teacher of the spiritual philosophy.

In the context of this study we are concerned neither with the instructor in practical skills nor with the exponent of academic systems of thought, but with the guru as *Mahātmā* (great soul), sage, saint, or seer, who on the basis of this experience teaches the way to transcendental knowledge both by exposition and demonstration. Such exposition is called his 'philosophy', but the Sanskrit word translated as philosophy has different connotations for the Hindu from those currently associated with European philosophy:

Philosophy ... is a word which formerly had a much wider significance than it now has and included under the heading 'natural Philosophy' all of what we now call science. Gradually, however, the term has become limited to speculative reasoning about the ultimate nature of the universe in its various aspects. It is true that a follower of Hegel would give quite a different answer to the question 'What is Philosophy?' to that which will be given by, say, Bertrand Russell, but in a general way we may say that philosophy means, in Europe, speculative reasoning about the universe, either on a basis of accepted principles *a priori* held to be valid, or on a basis of observed facts, and is an attempt to arrive at an understanding of the universe through the use of the discursive reason.

The classical systems of Indian philosophy, however, start on a different basis altogether. The Sanskrit word which is usually translated as 'philosophy' is *darshanam* which literally means 'seeing' and, in fact, the classical Indian philosophies start neither from *a priori* principles nor from observed facts, as usually understood, but from some transcendental experience in which the truth about the nature of the universe is directly perceived. What is usually called philosophy is an attempt to give a coherent and logical account of the world as thus perceived in such terms as shall be both intelligible and convincing to a given hearer with his own particular temperament. It is a rational demonstration of the truth seen by the original *rīshī* or 'seer', and with it is taught a practical method by practising which the pupil may gradually acquire for himself a realisation of the truths so demonstrated.

It is this claim to direct experience on the part of the teacher and to the possibility of such for the pupil that makes the widest gulf between the Indian and European systems. In the latter, no hope is held out that the pupil will ever arrive at more than an intellectual understanding of the truth. Whether one believes with Berkeley that a spade is an idea in the mind of God, or with McTaggart that it is a colony of souls, the practical result is just the same as if one believed with the crassest realist that it is just a spade. Whereas, when the author of the *Bhagavad-Gītā* says that 'all is *Vāsudeva*' (the indwelling spirit), he is saying something that he means his hearers to verify for themselves and on which they will be able to base their whole lives and outlook on the world.[1]

This brings us to the crux of the matter under discussion. If we confuse the seer's account of his direct experience with speculative philosophy; if we confuse mystical vision with mythological or archetypal dreams whose content comes from the collective unconscious, or think that the spiritual experience is a delusion of the disordered psyche in men seeking opiates against the stark reality of a life that ends in death, and that the death of the body is the death of the man; if we think that the guru, sage, saint, or seer is possessed by the 'old wise man' archetype or is suffering from some sort of megalomania; if, in fact, we are crass realists and believe that everything is just what it is, without reason, cause, purpose, or direction, and that anything perceived without the use of the physical organs of sense and without the brain is, *ipso facto*, hallucination, then we are blind, proud, stubborn, opinionated fools, lacking intelligence to inquire into obvious mysteries, lacking humility to suspect that others might conceivably be able to perceive facts which we are incapable of perceiving, and lacking courage to free our minds from the fixed opinions with which we screen ourselves from the impact of reality.

The experiential nature of the seer's perceptions does not challenge our philosophical ideas so much as it does the values on which we base our lives—not our pious values, but the ones like self-opinion and the importance given to feeling comfortable. It makes it easier for us if we can titillate our minds with his philosophy, give it a name, call it 'pessimistic' or 'obscure', and turn with relief to our Sunday newspapers which tell us how science is solving the world's problems, a science which regards even the commonplace ghost as 'unproven'.

This common and quite inadequate attitude is peculiar to the 'modern' world—the world most benefited by science and so most vulnerable to the materialism that goes with it. It seems, curiously enough, to represent a shadow cast by one of science's real contributions to the progress of the human race. By establishing the validity of experimental pragmatism when applied to the data of sense experience, science, as a historical movement, won its battle against religious prejudice and mythological thinking. Its antagonist was not merely the vested interest of institutionalized religion, which for centuries had cramped and distorted the development of thought, but the whole mode of primitive, mythological, and superstitious thought, the kind of thought which is still common in peasant and tribal societies. The development and diffusion of the scientific method, with its repercussions on philosophy, has had the effect of making our thought less mythical in content, more practical. We ask real questions and demand real answers. Our minds do not stop at the emotional resonances of the religious 'God', by whatever name called, but we ask, if we ask at all, for direct experience of the being or state of being which for millennia has been symbolized by the word 'God', even if we now call it the nature of reality.

We need not pride ourselves on this new ability. Men have sought and obtained such direct experience since there have been men on earth, and they did not have the disadvantages either of our rationality which we oppose to the objectively undemonstrable spirit, or the mass of scientific knowledge which clouds our perception of the magical qualities in nature. What they saw, however, they interpreted in the terms in which they thought; and much of their terminology no longer seems valid to us, unless we voluntarily make the effort to translate their concrete symbols into our abstract ones. Some progress has been made in this direction by psychology in its discovery that the 'unconscioussed' psyche of sophisticated people often uses the same sorts of mythological terms and symbols as did earlier men. In other words, our capacity for direct appreciation of the mythological outlook may be dormant, but it is not dead.

It is inadequate to regard this development of the conscious mind merely as a veneer of sophistication over an essentially primitive psyche. The man who uses his conscious mind to penetrate and integrate the primitive elements in the psyche does not go wild or run off in search of the simple life in savage surroundings. He becomes more, not less, civilized; more, not less, human.

Though the unitive experience has frequently been gained by uneducated and relatively inarticulate men of peasant or tribal stock, it is by no means limited to them. Some of the greatest minds of their times have known it, and some of them have been highly articulate in their attempts to interpret it, even though a few, like St Thomas Aquinas, found their eminent articulateness stopped by it. However, their interpretations were made in relation to the then current extent of world knowledge and therefore fail to provide satisfaction to many people of the present day who may be genuinely distressed at their inability to reconcile the demonstrations of saints with the demonstrations of science. Even though neither demonstration factually invalidates the other, few men are able to perceive the discrepancies that exist between facts and their interpretation—discrepancies that frequently underlie this problem of reconciliation. This is particularly the case where such discrepancies are made apparent only on the evidence of powers of perception which the questioner has not yet learned to use.

Although the scientist's facts may be objectively demonstrable, he takes into account only that aspect of his subject of study which can be made perceptible to the organs of sense, and he excludes the inner or subtle aspects of form from his frame of reference—many scientists deny that there is any evidence for their existence. His conclusions as to the significance of his discoveries are therefore inadequate, and when he extrapolates from his conclusions onto areas which come outside the self-imposed limits of his discipline, they may be entirely false.

Lest it be thought that this criticism of the outlook engendered by the scientific method is unjust, one may cite a review by Claude Alvares of the book *Chance and Necessity* written by the Nobel laureate, Jacques Monod: 'All systems rooted in animism, all religions, all philosophies, exist outside objective, non-purposive knowledge, outside truth, and are strangers to and thus fundamentally hostile to science. ... Our philosophers might just as well be dismissed, as they are no longer valid interpreters of truth.'[2] In his true enthusiasm for an uncomforting view of life, Alvares exceeds his competency to pronounce on the role of science in its contributions to the totality of knowledge. There are vast areas of objective experience outside the range of the material sciences which are as much parts of 'truth' as is the nature of the sort of matter which scientists have so far investigated.

We have no reason to suppose that men in the past who wrote learned theses based on inadequate or faulty knowledge of the material world were less intelligent than men in the present who write equally learned theses based on less faulty knowledge. Similarly, we have no reason to suppose that the level of mystical experience attained to today is any different from the levels of such experience in the past. But, because the attention of mankind has been turned towards inquiries into the material structure of the universe, development of inquiry into its inner nature has lagged behind. We have not yet learned to adapt the intellectual disciplines gained through outer inquiry to the less easily marshalled phenomena of the inner worlds.

So great has been the prestige of the scientific method that it is only now, when men have begun to taste the bitter consequences of the materialism engendered by the scientific outlook, that there is a widely felt need for a reinterpretation of life's significance. When men turn away from the sterile evaluation of life which has arisen from the scientist's unjustifiable extrapolations from his discoveries, we may expect developments in this field which will lead to reassessment of mystical experience with corresponding effects on the social ethos. These will to some extent balance the degradation of the old religious ethos which occurred when scientific empiricism supplanted mythological and superstitious thinking. For this to be effective, however, such reinterpretation must be a renewal of the eternal values of life based on direct mystical experience. No syncretistic re-hash by an oecumenical council will achieve any real change. To such men of mystical experience, the present accumulation of knowledge about the material universe will be a basis of fact with which their equally factual perceptions must be reconciled.

The human psyche cannot indefinitely tolerate our adherence to any one-sided view which opposes its need for integrity. Knowledge of the material universe may continue to be amassed, but the striving for total knowledge which is implanted in the soul of man demands that knowledge of existence be balanced by knowledge of essence.

As world views stand at present, we are divided roughly into two groups, the scientific and the religious: the people who accept the scientific view that reality is physical and that all things can be reduced to objectively measurable systems of energy, and the people who accept the supremacy of consciousness, the autonomous nature of the psyche, and the evidence for a non-physical reality. Each group reviles the other on the assumption that if one is right the other is wrong. But both views are expressed by the same human psyche, and, though the progress of human thought is often marked, and even promoted, by swings between such extremes, the psyche itself demands the reconciliation of extremes and their fusion into the wholeness which it seeks.

This apparent digression into the joust between science and religion is in fact relevant to our subject because it bears on the world attitude towards the seer as guru or spiritual teacher, and so on his practical, as opposed to his actual, status. To people for whom reality is what is perceived through the physical organs of sense, the seer is an anachronism, a purveyor of antiquated illusions. To the ordinarily religious people, he is the saint of their particular religion who affirms what they believe to be the uniquely valid tenets of their faith.

The Indian attitude towards the Mahatma, sage, saint, or seer, is similar to that of any predominantly peasant society. From familiarity with the traditions of his culture, the Hindu already knows the Mahatma, and he may also know him in fact. Mystical experience is perhaps not so rare amongst peoples whose teachings and customs are infused with perception of the immanent divinity as it is amongst those whose up-bringing persuades them to view all things as grossly and solely material. The Hindu is taught that the Mahatma is the embodiment of the highest, world-transcending knowledge, and that the pursuit of such knowledge is the greatest and most noble aim in life, ordained from the beginning as the ultimate goal of all men. He therefore reveres the Mahatma with that awe in which admiration and fear are mingled. The Mahatma is admired as a numinous figure, endowed with magical powers, whose blessings are insurance against misfortune, and whose teachings promise what may be variously interpreted as escape from or transcendence of suffering. His knowledge is essential and thus superior to any knowledge of temporalia. He is emancipated from desire, liberated from rebirth, and he neither seeks gain nor fears loss. He is also feared because, united with the universal spirit, he stands outside caste, religion, race, and social restraints in general, and therefore appears as a threat to the householder whose security lies in conformity to the established order.

Each Mahatma who arises currently represents the eternal source of the *dharma* or code of sacred conduct, established by gods, sages, and prophets in time immemorial as the mundane counterpart to the divine harmony of being. In the setting of mythological thinking the Mahatma therefore appears as a recurrence of a timeless beginning and consequently tends to be identified with one of the eternal gods or one of the sages of mythic antiquity. For this reason, men of actual spiritual attainment, whose inner and outer experience has liberated them from the conditioned thinking of the societies in which they arose, are mythologized by their devotees and publicly represented in a fashion that conforms to a culturally recognized pattern. On the other hand, men with equally valid experience, but whose minds have not been freed from behavioural rules, themselves conform to those modes of behaviour which their culture prescribes for the seer.

What, then, is the seer in himself? To answer this question we have to turn to the statements made by seers in their attempts to interpret the 'ineffable' vision, and ourselves attempt an interpretation which is consonant with our present understanding of the world.

Consciousness and the desire to 'conscire' are inseparable. Being desires to be. But, prior to the manifestation of a universe, the diffused awareness which inheres in unmanifest being can have no particularized knowledge of the content of being. Driven by its own desire, and from the energies of its own desire, the self-nature of being first establishes itself as the separate units of the atomic world. On this basis it then elaborates forms as vehicles for progressively higher modes of conscious function. Finally, in its highest and most elaborate form, namely that of man, it achieves an intensity of individualized awareness through which it is capable of obtaining both particularized knowledge of qualities inherent to its nature and knowledge of its prior state of unity out of which all energy, all form, and all awareness have sprung.

Inherently associated with consciousness, the urge to self-knowledge is the driving force within the evolution of forms, driving all creatures along the pathways of desire and, finally, urging men to discard their separate selfhood and to complete the cycle of creation which begins and ends in unity, a unity which the *Vedānta* of Shankaracharya describes as 'not two' or inseparate. Having exteriorized its own qualities in the differentiated forms of the manifest universe, through its out-turned awareness in its creatures, the spirit first appreciates its qualities in existence and then, turning inwards through man, rediscovers its undifferentiated essence. The cycle is complete. The term of the evolution has been reached. The spiritual experience is thus the natural and logical term of the progressive movement within consciousness through which the divine endeavour achieves its goal. Through its creatures, individualized in its human vehicle, the universal consciousness achieves awareness of its own essential nature.

In the unitive vision the identity of the individual with the universal is experienced, and it is perceived that this identity encompasses all being as an eternally valid fact. It has not come into being with the seer's attainment to the vision, but simply is. What comes into being, or, more truly, is developed in the seer, is the seer's capacity to perceive the identity. In this context it seems meaningless to say that any individual man ever attains anything. The spirit raises its human vehicle out of its own being and, through this vehicle, achieves knowledge both of the qualities it has made manifest to itself and of the undifferentiated and unmanifest being within which all qualities inhere. Our life is its life; our awareness is its awareness; our desire to live, to experience, and to know, is its desire. And the motivation which urges men to turn inwards to self-discovery is the driving motive behind the whole universe, a motive that seems to be as inherent to the nature of being as is consciousness itself.

The seers of the unitive vision are established in the unmoving centre of the world wheel. The unchanging Real which is their essential being was before all things, is now, and ever shall be after all things have ceased to exist. Within it, past, present and future blend into an eternal simultaneity, a simultaneity which is unaffected by the Einsteinian denial of temporal simultaneity because the former has nothing to do with the 'simultaneous' perception of objects through the organs of sense. That in the seer which stands in the centre always stands there by virtue of its nature. The worlds of form 'evolve' not outwards but inwards, enclosing, encircling, or integrating the eternal stasis within the circle of time. It is thus that within his being the seer links eternity with time, links them with full and constant awareness that the sequence of temporal events floats within eternity.

It is in this context that the seer as guru or teacher most significantly comes within the purview of these studies on philosophy and time. He shows the road from time into eternity, drawing other men towards him, as if the eternal in him called to the eternal in others. He does not go forth with missionary zeal to persuade others of the validity of his message. He is his message. Like calls to like, and those who have grown close enough to the eternal in themselves are attracted to the eternal in him.

From the above discussion it should be clear why the Hindu identifies the guru with God, whether by guru he means the particular individual through whose attainment the spirit now stands in full and non-illusory knowledge of its own nature, or whether he means the as yet unrealized potential within each one of us through which, if realized, the spirit will achieve self-knowledge. The latter, like any other unintegrated content or capacity of the psyche, urges towards inclusion in the awareness of the individual; urges, guides, and, when integrated, is the capacity for knowledge of the self-nature both of man and the universe. Prior to its development, for practical purposes we do not possess it, and we therefore become aware of it only in another man in whom it is realized.

To say that it is the spirit that attains its own goal and not the individual may seem to deny significance to persons and personal values. It is true that undifferentiated universal consciousness is so diffused as to appear impersonal, but if personality was not rooted in it as an inherent quality, there could be no person in man. By person we are not referring to the negatively toned mask of behaviour characteristics denigrated when we speak of a personality cult. We refer to the warm and living actuality of the man behind the mask, the indwelling spirit of man. When this person, this indwelling spirit, develops his full capacities and achieves the knowledge of his essential nature, he is not less human, but fully human. His knowledge and experience of the actual unity of being calls forth the compassion which is the underlying 'feeling-togetherness' of a unity which has expressed itself in parts. When he turns his attention towards his source, the objects of physical sense perception melt and vanish before his vision, and he stands in that same timeless void in which past, present and future co-exist inseparately. This is the 'void' from which all worlds emerged, in which they stand, and into which they will be withdrawn, just as for the seer they are now withdrawn. When, however, he turns outwards and is again aware of the worlds of form, he looks out as an individual through the limitations of his psychosomatic complex, limited by the accidents of birth, education, intellectual capacity, life experience, cultural conditioning, and the extent of world knowledge at the time. Furthermore, the society in which he resides may exert pressure on him to modify his statements. In early Christian and Islamic societies seers have been martyred because the truths of their experience were considered heretical. Indian seers were more fortunate; the code restricts behaviour but permits the mind to range free. Behaviour contrary to the code is tolerated if the practitioner removes himself from the social order and lives as a wandering mendicant. While this liberality may have led to an undisciplined proliferation of views, not all of which were based on experience, it has also assisted in preserving interpretations of mystical experience which are relatively free from the distortions that stem from religious coercion. Such limitations restrict the seer's interpretation of what he sees inwardly, just as they also restrict our interpretations of what we see outwardly. Nevertheless, despite differences in interpretation, the underlying similarity of the mystical experience gained by many seers is abundantly clear.

An example of this sort of limitation can be given from the life of Black Elk, a seer of the North American Sioux, who obtained a remarkable vision as a boy of nine, but could do nothing with it until older men with similar experience helped him to express it in the form of a ritual consistent with other rituals of the same tribe.

Another example can be taken from an account of the life of the south Indian Mahatma, Sri Ramana Maharshi, who similarly achieved the mystical experience as a youth. After his attainment he remained silent in a cave for some years. During this period a group of educated devotees frequently met in the cave and would discuss Vedantic philosophy together while sitting in the Maharshi's presence. According to one of the men who took part in these discussions, when the Maharshi broke his silence and began teaching, he taught in precisely those terms of orthodox *Advaita* which he had heard discussed. Neither this man nor others, including the present writer, who had the fortune to meet the Maharshi could doubt the reality of his status. But it appears that as a boy he lacked a framework of intellectual concepts in which to express his experience, and found the *Advaita* philosophy satisfactory when he heard it discussed.

Whether ancient or modern, Hindu, Buddhist, Moslem, Christian, or anything else, liberated from social values or bound by them, it is to the seers of mystical experience that we owe whatever validity there is in the religious and philosophical teachings of the world. They have 'seen' the one, timeless truth on which depends the validity of all the lesser and contingent truths of this world.

The significance of a seer far transcends his place in his local culture, for by virtue of his experience of the transcendental unity from which the whole of being derives, he represents the goal of all religions, the term of the evolutionary process, and the perfection of man. He is not merely the saint of a particular religion. Such differences as may appear between the Hindu Mahatma, the Islamic *Wali*, the Buddhist *Arhat*, and the Christian saint are accidental to their essential identity. All seers are essentially one, and their vision is one. Different though their temporal forms may be, in themselves they are the same.

Thus, any seer's mythic status within the sacramental framework of his culture in no way affects either the validity of his mystical experience and metaphysical knowledge or his significance, both to his own society and to the world at large. In discussing his essential significance, we are therefore concerned with him as a universal, rather than a local phenomenon. He is not a mere visionary, in the sense of someone who is susceptible to having clairvoyant perception. He 'sees' and experiences the essential unity of all things, and he integrates this experience into his being so that, in some sense, he lives in and from a constant awareness of the unity. The fact of the unity is his message; and the method by which men may attain his integral perception is his teaching.

The seer is the exemplar of the goal of life because the universal desire for self-knowledge has achieved its goal through him. All men have the ability for such achievement in potential, for the simple reason that they are part of the universe; and man is the one part of the universe which can know the whole, and know by his own experience his identity with the whole of being. When, therefore, the seer teaches, he calls forth this hidden capacity in other men, in the same sense that a school-teacher leads forth (educates) the unrealized capacities of a child's intelligence. If devoid of intelligence, the child could not respond. In practice, children ordinarily respond only to teaching levels within the range normal to the capacities of their age-groups. Similarly, men respond to the seer by virtue of their having the same capacity in potential which the seer has made real, but the level at which they understand or interpret the teaching depends on the extent to which their potentialities have been developed.

The outer teaching 'rings true' to men when it resonates in harmony with their innate capacity to experience the truths expressed. But since relatively few men have done anything towards developing this capacity, and since the teachings both from within and from without frequently conflict with the social and religious codes imposed on them in childhood, few men, again, can recognize and respond to the true teaching. They are presented with statements of fact which to them have no factual basis, because they have not developed the capacity to see or feel the demonstration of their truth in the being of the teacher. Nevertheless, the capacity is there, and it is usually true that inner responses are stirred by outer teachings long before the student's capacity is developed to the point where it can be trusted as a source of direct guidance. To draw yet another parallel from education, it is usually only in post-graduate work that a student can be trusted to apply the methods he has learned to independent studies and so to demonstrate that he has made them his own.

In the *Vinaya Pitaka* it is told of the Buddha that immediately after the enlightenment he doubted the possibility of conveying his knowledge to others. Anyone achieving similar enlightenment in the present day might feel the same way. The shared and relatively stable continuity of objective reference which fills the field sensed by the physical organs is declared the only reality, while private experience is denigrated as subjective and, by opposition to the publicly shared experience, unreal and illusory. Few people accept the truth of A. N. Whitehead's statement: 'Apart from the experience of subjects, there is nothing, nothing, bare nothingness.'[3]

All experience is private to the observer. Even the world of everyday life is privately observed by each one of us, and our perceptions are correlated by agreement. It is only because the unitive experience is as yet relatively rare that it is not commonly accorded the same validity as that of the everyday world. To the seer it is so convincingly real that he needs no confirmation of its nature; and this reality has been verified by Mahatmas, saints, sages and seers, from time immemorial and continues to be so verified. Furthermore, in discovering our identity with the universal substratum of consciousness, we discover the metaphysical unity which makes possible any sharing of objective content. If there were a plurality of separate consciousnesses, no two such consciousnesses could share experience, for they could have no grounds on which to build the supposition, which is what supports the whole scientific outlook, that all men perceive phenomena in much the same way.

However, the mystical experience is not shared in quite the same manner. Even if two seers were to get together to compare notes, they would not expect to find similarity in any objective content there might have been to their experiences. They would look for agreement in the significance of the direct knowledge or symbolized knowledge obtained. The visual or other objective content often associated with the unitive experience is different for different people according to their type, their cultural background, and the religious and philosophical ideas they have assimilated. This variation is inherent to the situation. In a realm where there is no form, any objective appearances are symbolic representations of the meaning with which the experience is filled; and these symbols are in some sense called forth or projected from the nature of the experiencing subject. They are symbolic interpretations of the essentially formless meaning within unmanifest being.

The problem becomes deeper when the seer returns to the 'normal' waking state and attempts to interpret what his nature has already shown as a symbolic interpretation. While the latter has at least the validity of what C. G. Jung calls the archetypes of the collective unconscious, the validity of the seer's reinterpretation will depend on the individual's mental capacity to handle the symbols of transcendental concepts and the adequacy of the philosophical systems with which he is familiar and to which he can relate his perceptions.

This may lead us to bark our shins on a stumbling-block, for we are apt to assume that spiritual teachers, especially those who teach from experience, express the unadulterated truth, whereas we now discover that what they express are interpretations of the truth. We need to come to terms with the fact that all statements about the transcendent unmanifest are necessarily symbolic interpretations, and that the use of a vocabulary of abstract terms is no less a symbolic interpretation than the use of concrete symbols, nor does it necessarily express more truth.

In fact, limitations to the seer's intellectual capacity and his ability to express himself coherently may present more problems for the disciple than for the seer. The seer may often rest content with a vocabulary that is meaningful to him. When a Hindu seer speaks of the *brahman*, he knows what it means to him in terms of experience and may not attempt to correct a listener who thinks he is talking about a specifically Hindu concept. On the other hand, we hear of a Zen master attempting a special study of philosophy in order to help his pupil.[4]

What the seer may say in his interpretation of the unitive experience is truly of less significance than what he is, for what he is validates what he says. His status is neither dependent on his learning, nor even on his capacity to articulate his experience. For anyone who has eyes to see, he demonstrates the truth in his being. The problem for the unenlightened man is how to perceive what the seer demonstrates. This is the problem that may face anyone seeking a teacher of the method by which the unitive experience may be obtained. How is one to distinguish between the charlatan, the scholar, the seer of distorted psychic vision, and the true seer?

All religious systems consider it statutory that their spiritual disciplines should be performed under the guidance of an approved teacher. Hinduism differs only in that it assumes the guru to be a true seer who has experienced his identity with the universal spirit; and its code of conduct prescribes an attitude towards any person accepted as a guru which equates him with the spirit. Where such rules are not framed solely in the interests of institutionalized religion, the principle behind them is valid. Were it not for the constant reassurance of the man who has been on this road before us, who stands by our side, affirming, encouraging and goading, which of us would dare, like the heroes of romance, to lose ourselves in order to find ourselves? And in a modern, non-sacramental society, men need reassurance that there really is something to be discovered and that the venture is not equivalent to madness. But, since the teacher is commonly within the socio-religious system, and since the motivation of the pupil is more often dissatisfaction with worldly life than a true inquiry, men publicly accepted as teachers of the spiritual way may factually be little more than worldly wise priests with stores of appropriate sayings from holy writ. One can find many benevolent, ascetic and dedicated men with great knowledge about the transcendent being, but with no direct knowledge of it. And there are professional teachers, both in India and elsewhere, who openly claim a status they do not possess, for the saintly garb is no guarantee that its wearer does not lie.

Hindu texts give descriptions of the true guru's distinguishing marks, some of which would seem to restrict the status to certain ethnic groups with particular physical characteristics, while others are applicable to a traditionalistic society where a specific dress and way of life may be laid down for anyone obtaining mystical vision. But since there are many levels of mystical vision, and since people without such vision may also follow the same way of life, there are factually no external signs by which the true teacher may be distinguished from the partial or the false.

The one reliable guide is 'the guru residing in the heart'. But we can rely on the voice of this teacher only when we learn to distinguish his 'voice' from the many voices of the as yet unintegrated psyche. Unless we have begun the task of awakening in our own hearts the capacity for the experience we seek, our hearts will not respond with recognition, and we shall be incapable of recognizing whether a man of apparent saintliness is speaking from learning or from experience. In our worldly professions we develop a fine sense for whether or not a man knows what he is talking about. We have to develop the same power of discrimination in this field.

Outwardly the seer is a man, waking and sleeping, eating and defecating, like anyone else. While one expects the emotional maturity of a man who has come to terms with his own nature, his mental capacities may or may not be above the average. If he teaches, he may say nothing which has not been said before and could not be learned from books. No external characteristic, no sensible sign, and no super-normal phenomenon necessarily differentiates him from the ordinary man. Outwardly he is ordinary. Inwardly he is exceptional because he has developed an actual capacity which in most of us is still unrealized potential. The development or integration of that potential capacity for unitive awareness constitutes the final step in the evolutionary progress of man. It is the one remaining power of being whose integration within the human soul produces the whole, complete, or perfected man.

Our aim, however, is not to win the capacity to distinguish true seers from false, but to follow the seer's path inwards and see for ourselves whether our experience corroborates his. We need give no name to the experience. We need neither philosophical hypothesis, scientific theory, nor religious formulation. All we need is the resolute courage to turn inwards to find direct experience of the roots of being, or, if we prefer to put it this way, to see whether our being has experienceable roots.

If we are at all intelligent, we do not need philosophical proof that God's existence can neither be proved nor disproved. The self-evident fact both of the universe and of our capacity to perceive its phenomenal appearances should be sufficient evidence to support any direct investigation into the source of the awareness by which we perceive. And since the investigation is of awareness and not of the material correlates of awareness, our sole course is to turn within ourselves. Our body forms the walls of our laboratory; our feelings, the power by which we operate; our dreams and visions, the data of the process; our thoughts, the regulators and channels of power; and that which is beyond thought, the awareness of being aware, becomes both subject and object of the investigation.

If we do not know that we are repeating an inquiry that has been successfully accomplished by many men before us, we shall be liable to stop at the first obstacle and come out with something like Descartes' *cogito ergo sum*. This barrier of the thinking mind is the first bottle-neck, corresponding to those between waking and sleeping, and between embodied living and disembodied living, rites of passage through which each man must pass alone, momentarily unsupported even by knowledge of his own identity.

No purpose would be served by adding one more to the many descriptions of the way, because the differences in individuals result in such varying approaches that we are all, in some sense, pioneers. Other men have blazed trails through the jungles of their minds, but we have never before travelled this road through our own. And if we find features in our country corresponding to features in theirs, it is not the fact that we have read or heard about this that gives it its validity; there is simply the pleasure of a shared experience. For instance, when the mind is truly stilled, the quality of the experience shows us why the author of the *Srimad Bhāgavata* describes the streams of *Braja* as 'clear as a Yogi's mind'.

We need only to follow our noses—inwards—in order to arrive at a confirmatory experience of man's essentially immaterial nature which carries such overwhelming conviction that material assessments of man are seen as sheer ignorance. But the disciplines we follow must be strict. If we waste our energies on temporalia, we have no one but ourselves to blame if our inquiry ends where it began.

We must expect that many elements in our natures will, like the crew of Columbus, rebel at our journey over the edge of the world. And we must expect that any conviction we find will be shared only with others who have braved the same crossing. We become like people with an extra sense whose data can be shared only amongst themselves. Those who lack the sense reject the data. We shall be called self-deluded, and our experience called hallucination. Yet he who knows his essential unity with all things can never be essentially alone.

Encompassing coincidence and what C. G. Jung called synchronicity, there is a harmony of being which brings it about that, when we search for the answer to our question in the one place where it can be found, namely in our individual focus of the universal awareness, outer correspondences are activated and come our way: books, people and other events meaningfully related to our search. This does not imply that we need never make effort to gain information. Opportunities are offered which, if taken, bring results appropriate to our needs. People looking for the teacher in the wrong place are easily duped by charlatans who claim them as their disciples.

Even if our fortune brings us before a man in whom we sense the presence of genuine knowledge and experience, we have no means of assessing his actual status and would be wasting our time were we to make the attempt. It often happens that our hearts and minds respond to the man who is only so far ahead as to be still in touch with us. We may draw inspiration from the presence of men who have gained the supreme knowledge, but the best guide in practice is often the man who is still working on or has just conquered the problems we are beginning to face.

Note that we speak of the heart and of feeling, and not of the mind alone. The experiential knowledge we seek is not a cognitive abstract, nor is it something we can get, look at, and file away for reference. By its very nature, the search for the whole demands wholeness in us who seek, and feelings are an important and often neglected part of the whole.

If we place ourselves under the guidance of any teacher of the way, we are not asking for religious or philosophical instruction, though both may be included in the discipline he imposes on us. We wish to follow him on his inner journey, share his experience, and, like him, discover the unitary consciousness within which we, together with all beings, are sustained.

The obstacles we shall encounter are hidden in our own natures behind those self-assertive demands and opinions by which the immature—and all men are immature who have not achieved wholeness of being—attempt to conceal their inadequacies of character. They are obstacles, because anything which drives us to assert our individuality thereby separates us from perception of the transcendental unity. And any fear of self-loss which makes us cling to externalities for reassurance, by asserting the importance of the outer universe prevents perception of the inner. If, therefore, the teacher produces emotionally charged situations which involve us, we must submit to them, for by using his own emotions he demonstrates to us the inadequacy of ours.

We shall attempt to protect ourselves from perception of our immaturities by permitting our own faults to be projected onto the teacher. If we fail to recognize what is happening, we shall eventually become so dissatisfied as to go off in search of a 'better' man, and, as a person who has failed in one marriage is apt to fail in another for the same reasons, we shall be liable to repeat the process of an initial enthusiasm followed by discontent when we balk at facing the same faults.

The principles outlined above can clearly be applied to ordinary life experience. We should be able to see that the true teacher is life itself. If we can accept both rational criticism of our faults and emotional and seemingly irrational demonstrations of them from a man we respect and love, without self-defence, self-justification, or projecting our own faults onto him, then we should be able to accept them from anyone. Life produces real situations which the teacher imitates. The teacher weighs the ties of dedication, loyalty and affection against the emotional strains he subjects us to, and he may take compensatory steps to offset shocks that prove too severe. Life appears to operate more impersonally. But if we examine the shocks life imposes on us and we see what lessons we have or should have learned from them, we can often perceive how they have acted on us towards producing maturity of being. However, since most men's views of maturity stop at a superficial adaptation to the practical affairs of living, it frequently happens that we cannot begin to see the teacher in life itself until we have gained a more adequate view from a human teacher. However, whether or not anyone applies this technique to us, it is in the midst of emotional turmoil that we can most easily recognize the one part of our being which is essentially calm in its own nature.

This attitude towards life experience is not an egotistical illusion arising from the idea either that the whole world runs round 'me', or that the universe is designed to benefit the human race. The universe is divine. The divine awareness seeks its own fulfilment through itself and finds it through the human vehicle in which that awareness achieves its highest intensity. In order to bring it to its highest intensity, the mental-emotional complex of the vehicle must be matured. Life experience acts towards maturity.

We have stressed the importance of the unitive experience, because its reality is the sole guarantee for the significance of our enterprise. But if we set out in search of 'an experience', we shall vitiate our efforts because such an attitude is egotistic and therefore separative. It is true that some people have an overwhelming vision of the unity, sometimes without much apparent effort, while others achieve it, if at all, only after years of intense struggle. But the former may take many years to assimilate and integrate what they have seen, while the latter may integrate each step as they take it. Admitting the reality of the experience and the maturity and integrity that it demands of us, how are we to interpret its significance?

Hindu philosophy, as commonly understood, places such importance on the unitive experience that the highest goal of man is represented as self-loss of the individual within the universal. However, the equally Indian philosophy of Buddhism, while denying the survival of anything that can be humanly conceived of as personal characteristics, in its Bodhisattva doctrine gives as an alternative goal that of the individual who, refusing complete dissolution in the transcendent unity, retains an individual spiritual existence out of compassion for the suffering of all beings.

And whereas the *Advaita Vedānta* of Shankaracharya, commonly considered the most representative and influential form of Hindu philosophy, stresses the unmanifest and transcendent reality to the detriment of the manifest universe which it declares illusory, Nāgārjuna of the Mādhyamika school of Buddhism declares: 'There is no difference at all between *Nirvāna* and *Samsāra*'—meaning thereby that the unmanifest absolute and the worlds of form are of the same nature.

We are not concerned to evaluate these views either in support of or in opposition to the religious systems which claim them as their own. They are not given as examples of religious thought, but in order to demonstrate the different interpretations given by different men of what was fundamentally the same experience. Both views represent modes of apperception of the one truth. And we must have the honesty to admit that the ultimate truth may not be susceptible of any one definitive evaluation. All that can be said with certainty is that it is the underlying unity of being which supports all values. This could be formulated in the words of an elderly Indian sufi who, when asked 'What is truth?', replied: 'That is very simple. What makes one out of two is true. What makes two out of one is false.'

From the attitudes taken by seers towards their own perceptions, it would appear that we are here not so much dealing with two different interpretations of one truth, one of which must be more true than the other, as we are with the concept of there being two distinct interpretations, each of which, if accepted, leads to different results. All seers agree that the unity can be entered and that the individual who enters it is lost for ever. From the accounts of seers who have approached that state, such annihilation would appear to be utterly blissful. Whether blissful or not, however, annihilation in the unmanifest being from which all things arose is the inescapable end of all existence. And the fact that the divine awareness joys in uniting itself with itself and includes the individual in its joy may not, so long as we have not reached that state of transcendental being, appear to be adequate compensation for our involvement in a seemingly pointless system of repetitious self-creation followed by self-destruction.

Though each individual who achieves such utter annihilation presumably has his moment when he feels participation in the cosmic fulfilment which is attained through him, this achievement is obtained at the cost of its immediate obliteration. Nothing remains of the individual, and the universal awareness loses its focus of individual experience through which it achieved the self-knowledge it seeks. Both the achievement and the world-negating thrust which supported the achievement appear to make nonsense of the seemingly tremendous effort which has raised existent being to the point where individuals can attain such intensity of awareness and such intensity of directed effort that both world and individual can be neglected.

Though all seers accept the reality of this annihilation, not all of them describe it as the only or even as the highest course. The Bodhisattva doctrine, mentioned above, which teaches that men may halt on the edge of the ultimate unity for the benefit of suffering humanity, has its parallels in all religious systems. This doctrine places annihilation as the highest goal which brings the greatest bliss to the participant, and it represents restraint from this goal as a sacrifice, motivated by compassion, a compassion that is no sentimental 'do-goodism', but which derives from actual perception of the essential unity of all things. However, some seers go so far as to denigrate annihilation in the sense of self-obliteration in the infinite, calling it selfish bliss, and prefer restraint as the highest goal. The rationale for such preference is simply this: through the individual attainment the universal spirit achieves its goal of self-knowledge. If such a man passes into utter annihilation within universal being, the goal is lost at the moment it is found and the universal effort is vitiated. If he refuses annihilation, the spirit retains its self-knowledge in his being. Just as the unmanifest spirit needs its manifest form in order to know its inherent qualities, so it needs its essential knowledge to be held in individual forms if it is to be retained and the two aspects of knowledge fused.

In evaluating such restraint as the highest goal, the seer is not cushioning either himself or us from the apparent starkness of annihilation. Nor can we impute to him a wish for egotistic self-preservation. Only self-transcendence can achieve either goal, and only the self-restraint natural to love can achieve the restraint necessary to refuse the unifying pull of the transcendent being which is experienced as its love-desire. Indeed, such restraint is not the sacrifice of the highest goal, but the final sacrifice of ego-motivation. The seer who achieves this state is no longer in any sense a separate individual. The light of the universal awareness shines, unobstructed, through the vehicle of its own form.

This dual interpretation of the goal possibly accounts for there being two recognizably different attitudes taken by seers towards the world: those who sacrifice the bliss of self-annihilation commonly accept the bonds and responsibilities of affection here, while those drawn to annihilation commonly reject them.

Myth, folklore, the teachings of religions, the living experience of men engaged in this work, and the statements of seers themselves, all affirm the real existence of such great beings who, long after their physical deaths, continue to live and to guide others to their goal. No one who has ventured far along this path will dismiss these figures as illusory images of the 'old wise man' archetype, as, apparently, would the school of C. G. Jung. Nor, for that matter, will anyone who has met an embodied seer dismiss him as being possessed by such an archetype.

Anyone who undertakes this work in all seriousness will find himself compelled to accept the objective reality of subtle planes of being. They constitute both the realms of after-death existence and the subtle components of physically embodied existence. Anyone who withdraws his attention from the organs of sense inevitably finds his powers of sense registering phenomena of another sort. And though the data supplied by the sense powers are unreliable, in that they tend to be mixed with the projected content of the seer's 'unconscious' psyche, analytic sorting of such data brings to light sufficient evidence for the same sharing of experience as constitutes the criterion of reality for the world perceived through the physical organs. Furthermore, anyone who takes the trouble to do so can gain sufficient experience of the physical and psychological effects produced by disembodied men impinging on this physical world to convince him that the shared experience of subtle states of being is neither collective hallucination nor mere experience of the collective unconscious.

The seer has passed through these realms and, though he may discourage us from egotistic fantasies of development and exploitation of psychic powers, he knows the value of paranormal experiences which shake our habitually materialistic outlook. He introduces us to a path on which, if nothing unusual ever happens, we may justly doubt either our own progress or the efficacy of the method we are practising.

The effect of mystical experience is not limited to the individual to whom it comes. The individual becomes the source of a world view which affirms that life is meaningful and purposeful only when the direction of events along the dimension of time is related to their significance in the dimension of eternity.

We have spoken of the seer as the exemplar of the goal of life and as the sum of human perfection, yet so described his characteristics as to include inarticulate men of average intelligence. Though we can easily accept that there may be human qualities of greater general significance than intellectual brilliance, it is hard to accept a view of human perfection in which there is no visible lustre. It is true that the human eye is a window into the soul which gives us a measure of the soul's radiance, but it is the tongue that reveals or betrays our highest human qualities. To the uninitiated a rough diamond may pass unnoticed. Let but one face be chipped, and its qualities as a gem become apparent. Not until many facets are skilfully cut and polished is its full beauty and brilliance manifested. Nevertheless, one recognizable diamond, even uncut, is of more value than highly polished glass. And one real seer, even if inarticulate, is of more value than all the highly polished prelates of the world.

The above analogy suggests that there may be seers with many or all of the facets of their personalities highly developed. And so there are. But one must beware of fantasies on this subject. Before hunting a presumptive seer with shining facets one must learn to distinguish diamond from glass.

When the divine awareness achieves knowledge of the divine self-nature, it achieves the term of the movement by which it evolves itself out of itself. By the fact of his being, the human teacher or guru exemplifies the term of the evolution and, by virtue of his example, arouses and activates his counterpart within our hearts.

The guru lives in us as the urge to wholeness. We see him in men who have found the whole or part of that whole. We see him in the whole of life that grows towards wholeness. We find him in that unity which contains the whole.

The wonder with which we view the guru's path as it unfolds before us, the spontaneous respect we feel for him whose calm eyes affirm that our turmoil can be calmed, the love that wells up in our hearts for him who supports us in our moments of despair, and our overflowing gratitude to him who gives us so much of his knowledge, of his guidance and, above all, of himself—all these are natural responses of the human heart towards the man whose very being declares the truth of his attainment and the factual efficacy of his teaching. By being what he is, he has shown us the way, for he is in himself the demonstration of an objectively undemonstrable attainment.

Praise to the Guru who showed me the impartite Spirit which pervades all animate and inanimate beings.
Praise to the Guru who cured my blind ignorance with the eye-salve of knowledge.
Praise to the Guru who is Creator, Preserver and Destroyer, and is even the Supreme Being himself.[5]

NOTES

1. R. H. Nixon (Sri Krishna Prem), 'An Outline Sketch of Buddhist Philosophy', *c.* 1925. (No record is available of the journal in which the article was published.)
2. *The Times of India*, Sunday, 15 October 1972.
3. A. N. Whitehead, *Process and Reality*, London, Macmillan, 1967.
4. Eugen Herrigel, *Zen in the Art of Archery*, London, Routledge & Kegan Paul, 1953.
5. From the Sanskrit *Hymn to the Guru*.

THE LEADER

Arnold Toynbee

The epithet 'great'—as applied, for instance, to Peter the Emperor and to Catherine the Empress of Russia—implies a recognition of the truth that human beings are endowed by nature with different degrees of ability. This fact has been acknowledged most readily in the sphere of political activity, but it can be recognized in any sphere. The medieval Western philosopher Albert was called 'the Great', and his right to the title is more evident than the right of any medieval Christian emperor or king. But many men and women who have been truly great have lived and died without being given the epithet. They have remained unrecognized because they have not had the opportunity of exercising their exceptional natural gifts and displaying these to their fellow human beings. This is one of the themes of a famous elegiac poem by the eighteenth-century English poet Gray.

The term 'historical figure' is more illuminating than the term 'great man', because it concedes that the greatest of great men needs to be born at some particularly favourable moment in order to bring his potential greatness to fruition. The word 'leader' is still more apt, because it declares unambiguously that there is a relation between a leader and his followers, and that a potentially 'great man' has to establish this relation between himself and his fellow human beings in order to convert his potentiality into an accomplished and applauded fact. The word 'leader' implies the existence of another party, and an interaction between 'the leader' and 'the led'.

The use of the word 'leader' is therefore not only convenient; it is indispensable, in spite of the temporary ill-odour of this word in the Italian and German languages, as a result of the careers of leaders who, in our time, have betrayed the trust that their followers have placed in them. However, we may expect to see the words *duce* and *Führer* rehabilitated by future leaders who will be worthy of this title, and in the Greek language the corresponding word has never forfeited its auspicious connotation. 'Leader of souls' (*psychopompos*) was one of the titles of the pre-Christian Greek god Hermes. 'Pathfinder' (*hodiegetria*) is one of the Greek titles of the Virgin Mary.

A potential leader needs responsive followers; a nation that is in desperate straits needs a charismatic leader. This mutual need is particularly manifest in critical military and political situations. The British people could hardly have escaped being defeated in the Second World War if it had not found a great man, Churchill, to lead it in its hour of need. But Churchill, on his side, could not have played his part if the British people had not been in the mood to accept Churchill's offer of 'blood, sweat, and tears' as the necessary conditions for pursuing the goal of victory. In 1940, this goal seemed to be hardly within either Churchill's or Britain's reach. Nevertheless, in unison they did achieve victory, and they did this by giving each other the mutual support that was the key to success. Twenty years before 1940 the Turks had been in an equally desperate plight, and they too had won a victory by responding to the appeal of a great leader, Atatürk, to follow him in fighting against apparently hopeless odds. The Turks' Byzantine Greek predecessors had responded similarly to the Emperor Heraclius' call. The coins minted by Heraclius during that crisis in the Byzantine Empire's history bear the moving inscription: 'God help the Romans!' In this case, the leader himself lost hope at the most critical moment, and the people whom he was leading took over the leadership themselves. They reanimated their leader's courage and inspired him to persevere until he had led them to victory. In this instance, the reciprocity between the leader and his followers is particularly conspicuous.

More than a thousand years before Heraclius' time, the Athenians had saved Greece from being conquered by the Persian Empire thanks to Themistocles' leadership and the Athenian people's acceptance of it. When a lucky strike of silver was made in Attica, Themistocles persuaded his fellow countrymen to spend this trove of precious metal on building a fleet instead of distributing the silver among the citizens of Athens as a bonus. And then, when the invasion came, Themistocles persuaded the Athenians to evacuate the mainland and to leave it to be devastated by the invader in order to gain the opportunity of winning a decisive naval victory with their newly built fleet in the straits between the mainland and the island of Salamis.

A potential leader needs his opportunity and, conversely, a public crisis calls for a great man to inspire the people to rise to the occasion.

Someone has written an imaginary history of the modern world in which the French Revolution and the Napoleonic Wars do not occur. The author of this *jeu d'esprit* records the deaths, in 1852, of two octogenarian military officers, a British major-general and a French captain. The author professes to have found a short obituary notice of the British officer, Arthur Wellesley, in the London *Times*. Wellesley had distinguished himself in some Indian campaigns in the early years of the nineteenth century, but, when Britain had completed her conquest of India, there was no more opportunity for Major-General Wellesley to win laurels. The French captain had never had a chance of seeing active service. He had reached retirement age before 1830, the year in which France had started to conquer Algeria, and this (according to the writer's imaginary reconstruction of history) had been France's first war since her intervention in the American war of independence, more than half a century earlier. This French artillery officer was too obscure for his death to have been reported in the newspapers. The author professes to have found by chance a notice of it in the parish church's register of deaths in a small town on the French Riviera in which this retired officer had spent his last years. The French officer had an Italian-sounding name, Napoleone Buonaparte, but there was no record of his birthplace.

Conversely, the needs of a great country in a great crisis cannot confer greatness on an undistinguished person. A little man from the little island Belbina is reported to have said to Themistocles: 'You could never have done what you have done if your fatherland had been mine, Belbina, instead of being Athens.' The little man was trying to depreciate Themistocles' achievement by attributing it to the greatness of Athens exclusively. Themistocles crushed him by retorting: 'You are right, but it is also true that *you* would never have done what I have done if your fatherland had been Athens, instead of being Belbina.' Athens' critical situation in 480 B.C. cried out for a leader, and Themistocles played the necessary role, but this could never have been played by the little man who exposed himself to Themistocles' crushing rejoinder.

This historic conjunction between a public crisis and a great leader can be illustrated by a number of other celebrated examples.

The first Chinese Emperor and, two centuries later, the first Roman Emperor, Augustus, each saved a whole society from destruction by giving it a strong unitary government. They each had been given the same opportunity. The society that each of them saved had formerly been divided up politically among a large number of local states that had been perpetually at war with each other. On each occasion, the politically divided society was in desperate need of peace, and it was given peace, just in time, by a unifying Emperor. Ch'in Shih Hwang-ti's greatness and Augustus' greatness was offered the chance to manifest itself by the situation of the great man's society in the great man's time. We have to take account of this situation in order to understand why Augustus and Shih Hwang-ti were able to do what each of them did.

Saint Francis of Assisi is the greatest soul that has made its appearance in Western Christendom so far, but Saint Francis, too, cannot be understood without taking account of his circumstances. Western Christendom in Saint Francis' day was becoming rich. Saint Francis' father was a successful wholesale cloth merchant. Within Western Christendom, the Church and, within the Church, the Papacy, was becoming powerful and was abusing its power. Consequently the Papacy and the Church as a whole were falling into disrepute. The criticism of them was taking the practical form of dissenting religious movements. There was a proliferation of Christian sects that were dedicated to poverty, as a protest against the ecclesiastical 'establishment's' greed for money and expenditure of its revenues on sumptuous living. A rival religion to Christianity, Catharism, was spreading from Bulgaria to Lombardy and Languedoc. The top-ranking Cathars were ascetics.

Saint Francis repudiated his father's wealth, espoused 'the Lady Poverty', and persuaded Pope Innocent III and his cardinals to welcome the Franciscans as saviours of the Christian Church instead of crushing them as heretics. If, in Saint Francis' time, the Western Christian Church had not been corrupt and discredited, and if Innocent III's advisers had not perceived that Saint Francis was the great man who was needed for saving the Church from destruction, Saint Francis might have been written off as a 'hippy' by his fellow townsmen in Assisi and might have lived and died unknown to the rest of the World.

Charles Darwin demonstrated, in a book, *The Origin of Species*, published in 1859, that the innumerable species of living organisms, including man, could have evolved without any need for the personal action of a creator god. If Darwin had published his revolutionary book in Morocco in 1859 or in Britain in 1459, he would have been put to death as an atheist. Even in Britain in 1859 *The Origin of Species* evoked violent opposition. It eventually won general acceptance in the Western world because, for about two centuries preceding the year 1859, Westerners had been becoming scientific-minded and sceptical about the dogmas of the Christian religion.

Bismarck was one of the most successful statesmen of the nineteenth-century Western world, and his success was obviously due in part to his personal qualities. He set himself limited aims and resisted the temptation to go beyond them when this was within his power. After having defeated the Habsburg Monarchy, Bismarck left it territorially intact. He was content with achieving his original objective, which had been to exclude Austria from having any further say in the affairs of other German States. Then, when he had defeated France, Bismarck resisted, as far as he could, the Prussian General Staff's demand to detach from France places (e.g. Belfort) whose inhabitants were French-speaking. Yet Bismarck would not have been able to unite the whole of Germany if he had been a contemporary of Frederick the Great. Bismarck was given the opportunity that he seized because, in the nineteenth century, the German people were ready at last for political unification.

The Germans had been humiliated by Napoleon's conquest of Germany, and they had realized that the reason why Germany had been at France's mercy militarily was because Germany had been divided politically among a number of separate states, no one of which was a match for France singly. Then, after the Napoleonic Wars, the Industrial Revolution had begun in Germany, and the Germans had realized that mechanized industry could not be developed effectively in a country that was not united politically. Before Bismarck entered on his political career, the local German states had already formed a customs union on the initiative of the Kingdom of Prussia, and a political union was the obvious next step. Finally, by the 1860s, the Kingdom of Prussia was able to provide an effective political and military nucleus for the unification of Germany thanks to the cumulative results of the work of Frederick the Great, Stein, and other eminent Prussian statesmen who had prepared the way for Bismarck. So Bismarck the statesman, like Darwin the scientist and like Francis the saint, did not live and work in a social vacuum. The time was ripe for his greatness to find scope.

The Prophet Muhammad looks, at first sight, like a great man who changed the course of mankind's history solely by his personal genius. He preached monotheism to a backward people living in an out-of-the-way corner, and he inspired his Arab fellow countrymen, whom he had converted to his new religion, Islam, to spread this religion, by force of arms, from Arabia to the adjoining civilized regions. Muhammad's career and its sequel have sometimes been cited as an instance of an historical event that was necessarily unpredictable because it is inexplicable. On closer inspection, however, it can be seen that a sharp-sighted observer could have perceived, by Muhammad's time, that there was now bound to be a revolutionary change in Arabia, and that this coming Arabian revolution was bound to affect other parts of the world.

By the year 632, Muhammad had unified Arabia politically, whereas the Roman and the Persian Empire, which were Arabia's next-door neighbours in the north, had exhausted themselves by waging against each other a long war that had been devastating but inconclusive. The two exhausted empires were an easy prey for the Arabs' united forces. The unifying force in Arabia was Muhammad's monotheistic religion, Islam. But, in Arabia, the ground had been prepared for monotheism, and the seeds had been sown, long before Muhammad arrived to reap the harvest. First Judaism, and then Christianity, had been seeping into Arabia from 'the Fertile Crescent' on the north and from Ethiopia on the south-west. The Arabs had become aware of not yet being, as their neighbours were, a people endowed with a holy scripture. In providing the Arabs with the *Qur'ān*, a sacred book in the Arabs' own language, Muhammad was meeting a need that, in his time, his fellow countrymen were feeling. So Muhammad's career turns out not to have been due solely to the greatness of the Prophet's own personality. His career is not inexplicable and was not unpredictable. In this case, too, the event that gave a new turn to history was a conjunction of greatness with opportunity.

Muhammad was great on both the religious and the political plane. If we now look at the careers of a great political leader, Napoleon, and a great religious leader, St Paul, we shall see that they, too, were fortunate in having been given their opportunity. At an earlier point in this essay, it has already been suggested that Napoleon might have lived and died in obscurity if the French Revolution had not occurred at a date at which Napoleon was a young officer in the French Army. But the French Revolution was not the only political event in Europe that played a part in making Napoleon's fortune. For several centuries before the outbreak of the French Revolution, France's neighbours, Germany and Italy, had been disunited politically and had therefore been at the mercy of France and other adjoining great powers. It has been noted already that Napoleon's conquest of Germany was one of the historical factors that gave Bismarck his opportunity later on in the nineteenth century. Napoleon built his empire by excluding Prussia and Austria from having any say in the affairs of the rest of Germany, and by excluding Austria from Italy as well. Napoleon attached to France the petty States among which both Italy and western Germany had been divided. Napoleon's success in making himself master, first of France, and then of Germany and Italy is explained by a contemporary event in France, the French Revolution, and by a situation in Italy and in Germany that was the outcome of a gradual development in the course of several centuries. Napoleon's career was not due solely to Napoleon's personal genius.

St Paul's achievement was as revolutionary as Napoleon's and as Muhammad's, and, like Muhammad's, it was permanent, in contrast to the ephemeralness of the Napoleonic Empire. But St Paul's career resembles both Muhammad's and Napoleon's in being a great man's seizure of a favourable occasion for giving scope for his genius. St Paul turned a Jewish sect into an oecumenical religion, and this was an amazing achievement, but St Paul could not have accomplished it if the time had not been ripe.

In the Roman Empire in St Paul's day, as in Arabia in Muhammad's day, people were feeling that their ancestral polytheistic religions were spiritually inadequate. They were hungry for a monotheistic religion, and consequently the local Jewish communities that had established themselves all round the Mediterranean basin had attracted non-Jewish adherents. These had embraced Jewish monotheism, but they had been repelled by the code of manners and customs that was obligatory for Jews, and they had continued to be semi-outsiders, not full members of the Jewish community. St Paul was unable to convince the Jews that Jesus was divine. For Jews, to believe in the divinity of any human being would have involved a repudiation of the fundamental tenet of Judaism. For non-Jews, on the other hand, the notion that a human being was a god was familiar and congenial. The most recently deified man, in a long series of deifications, was the Emperor Augustus, who had saved civilization from collapse at the western end of the Old World. Before Augustus, Alexander the Great and a number of his successors had been deified, and, before Alexander, Pharaohs of Egypt for at least 2,000 years. The non-Jewish adherents of Judaism were ready to believe that Jesus was God, and, on this condition, St Paul was able to admit them to membership in a non-Jewish Christian community without any obligation for them to be bound by the Jewish law. The previous history of both the Jews and the non-Jews in the countries round the Mediterranean Sea gave St Paul his opportunity.

Chandragupta Maurya united politically the greater part of the Indian subcontinent in the fourth century B.C. As an empire-builder, he was a man of genius, but the circumstances were also propitious for him. In the basin of the rivers Jumna and Ganges, political unification had been achieved already. In the basin of the River Indus, the local states had just been conquered by Alexander the Great, but, after Alexander's death, his successors were unable to maintain their hold on a region that was so remote geographically from the Macedonian conquerors' homeland. Chandragupta seized the chance to expel the Macedonian garrisons from the Indus basin and to use this territory as a base for conquering the already united Jumna–Ganges basin. If Chandragupta had had to start building his empire from the beginning, without finding ready-made foundations that had been laid by his predecessors, he might have failed to accomplish what he actually achieved.

The Muslim mystic al-Ghazālī induced the Islamic religious 'establishment' to accept mysticism as being compatible with Islamic orthodoxy. This diplomatic feat of al-Ghazālī's was remarkable, considering that, for the ecclesiastical authorities of a theistic religion, mysticism is bound to be suspect. Mysticism tends to break down the barrier between God and Man, but, for believers in a theistic religion, both God and Man are individual personalities, and, if they blend into a unity, the individuality and the personality of each of the two parties is lost. They have ceased to be two; they have become one.

For this reason, both Muslim and Christian mystics have sometimes been treated harshly. Before al-Ghazālī's time, the Muslim mystic al-Hallāj had been put to death for having declared that he was identical with the godhead. Why was al-Ghazālī able to reconcile the Islamic 'establishment' to mysticism? Al-Ghazālī succeeded, where al-Hallāj had failed, because, by al-Ghazālī's time, the Islamic World was falling into adversity, and therefore Muslims were now longing for the comfort and support of a more intimate relation between Man and God. They were yearning for a presentation of God in which He would be found to be less aloof and less forbidding than He had appeared to be in the original Islamic conception of him. Al-Ghazālī met these new spiritual needs. If he had been al-Hallāj's contemporary, he might have shared al-Hallāj's fate.

At an earlier stage of Islamic history, abū Muslim had engineered a political revolution which, like the Prophet Muhammad's, looks at first sight as if it had been abrupt. After only about three years of underground propaganda, abū Muslim was able to launch an overt revolutionary movement that overthrew the Umayyad dynasty. Abū Muslim was a gifted revolutionary leader and he was given the opportunity of starting a conflagration because there was tinder for him to ignite.

When a leader's followers go on strike, even the most highly gifted leader is powerless. Alexander the Great's sensational conquest of the huge Persian Empire was won thanks to his father Philip's bequest to him of the extremely efficient army that Philip had created. So long as this army carried out Alexander's orders, Alexander was invincible. When, at the right bank of the River Beas, the south-easternmost of the five rivers of the Punjab, the army refused to advance farther, Alexander had to submit and to retreat. Alexander's genius was given as much scope as the Macedonian army was willing to grant. Alexander would have lived and died as the ruler of a backward kingdom on the fringe of the Greek World if his father and predecessor, Philip, had not created an army and imposed peace on the warring city-states of the Greek World. The Persian Empire had been secure against invasion and conquest so long as Greece had remained politically disunited. If Alexander had been Philip's father, not his son, Alexander's immense military and political abilities would have been expended on local victories in minor border wars.

Alexander's actual achievement, and its limits, illustrate the importance of the role played by a great man's circumstances in deciding a great man's fortunes. This point is also illustrated by the well-known fact that a movement is sometimes launched, and a discovery is sometimes made, by more than one person simultaneously. Muhammad was the successful Prophet in Arabia in the seventh century of the Christian Era, but he was not unique; he had an unsuccessful rival, Maslamah. In the 1850s, two Englishmen, Charles Darwin and Alfred Russell Wallace, simultaneously reached the conclusion that the variety of species of living beings on the planet Earth could be accounted for without need of the hypothesis that the world had been created by a God who possessed the human-like capacity for entertaining and executing a purpose. In the 1860s, three Englishmen claimed credit for having discovered the source of the White Nile. These instances of simultaneity suggest that, in the interplay between genius and opportunity, it is opportunity that counts for most. In the seventh century the development of religion in Arabia had arrived at a stage at which it offered an opportunity for prophets of monotheism. In the 1850s the development of scientific knowledge and understanding in England had arrived at a stage at which it provided a clue for an explanation of the evolution of life. In the 1860s the opening-up of the interior of Africa by European explorers had arrived at a stage at which several of these could compete with each other for the prize of being the first European to reach the source of the White Nile.

The importance of timing is illustrated by cases of precocious failures and of posthumous successes.

In the history of the development of science in England, Roger Bacon was so far ahead of his time that his work was incomprehensible to his contemporaries and was therefore disregarded. Francis Bacon was near enough to the Promised Land to catch sight of it lying ahead of him, but he was not near enough to be able to set foot on it himself. But Francis Bacon's premonitions did inspire the men of science in the next generation to found the Royal Society in the year 1660, and, from that time onwards, the intellectual climate has been propitious enough to enable science to advance, not only continuously, but at an accelerating pace.

The penalty of precociousness can also be illustrated by examples in the field of politics. Ever since the political unification of China in 221 B.C., the Chinese people have been wrestling with the problem of providing for a unitary administration without saddling themselves with a civil service that exploits its power for the personal advantage of its members and their families. Early in the first century of the Christian Era, a first attempt to solve this problem was made by Wang Mang. In the second half of the eleventh century, a second attempt was made by Wang An-Shih. At last, by the mid-point of the twentieth century, the time had become ripe for a third attempt to provide for an effective administration of China without giving an opening for the rise of a new corporation of mandarins. The Chinese people had been suffering from mandarins for about 2,000 years, and they had suffered enough to move them now to make a supreme effort. This appears to have been the purpose of Chairman Mao's 'Cultural Revolution'.

In the history of the Roman Commonwealth, Gaius Gracchus had perceived, before he was elected to be one of the tribunes of the plebs for the year 123 B.C., that the formerly efficient Roman nobility had become unfit for governing an empire that now embraced the whole basin of the Mediterranean. If Gaius Gracchus had had a free hand, he would have converted his temporary tribuneship into a permanent dictatorship and would have given good government to Rome's citizens and subjects at the price of depriving the Roman oligarchy of its political privileges. However, Gaius Gracchus was before his time, so, like his elder brother Tiberius in 133 B.C., Gaius met with a violent end. A century later, Augustus achieved, with personal impunity, the political reforms which Gaius Gracchus had failed to accomplish. Augustus was successful because, by his generation, the time was ripe. An additional century of civil war and political turmoil had reconciled the Roman people to sacrificing their traditional republican régime in order to purchase the boon of orderly government.

The cases of posthumous success are no less striking. By Jesus' time, the prophets of Israel and Judah, who had been ignored or opposed in their own lifetimes, had come to be revered. Confucius resigned himself to becoming a teacher of moral and political philosophy as a *pis aller*, because he had failed to make a successful career for himself as a civil servant. After the political unification of China, Confucius' teachings were made canonical for the education and guidance of the Chinese Imperial civil service. Confucius had not foreseen the political unification of China, and probably he would have disapproved of it; for this was a revolutionary political change, and Confucius was conservative-minded. In the domain of biology, Mendel's record of his observations lay neglected for years until the science of genetics had progressed far enough for Mendel's successors to appreciate the significance of his work.

An indispensable condition for success in leadership is the gift of psychological insight and sympathy. The leader must carry his followers with him. If he fails to do that, his genius will be frustrated. Though Alexander the Great was idolized by his soldiers, he eventually demanded of them more than they were willing to give, and then even Alexander had to retreat. Impatient and imperious leaders have provoked opposition that has undone their work. Ch'in Shih Hwang-ti, the first unifier of China, rationalized and standardized the administration of the newly established Chinese Empire by such high-handed acts that, immediately after his death, there was a general revolt and a bout of anarchy and civil war. The second unifier of China, Han Liu P'ang, read correctly the lesson of his predecessor's failure. He hastened slowly. The work that had been done by Shih Hwang-ti in twelve years was re-done by Liu P'ang and by Liu P'ang's successors in the course of three-quarters of a century. Consequently their work endured.

Augustus' uncle and adoptive father, Julius Caesar, met with a violent end, like the two Gracchi brothers before him. Augustus was no match for Julius Caesar in point of genius, but Augustus proceeded deliberately with a tactfulness and a cautiousness that Julius Caesar would have disdained. Therefore Augustus' work lasted, after Julius Caesar's work had gone with the wind. An impatient and imperious temperament was also the ruin of the reformer Pharaoh Akhenaten. His high-handedness mortally offended the Egyptian ecclesiastical 'establishment', and he failed to win the support of the rest of the Egyptian people because of his haughty aloofness. Pope Gregory VII was a counterpart of Akhenaten. His militancy was counter-productive. Gregory VII's predecessor Leo IX had the art of disciplining members of the hierarchy without alienating them; Gregory VII's successor, Urban II, had the art of arousing popular enthusiasm. Gregory lacked both these gifts, and therefore his career ended in failure. Wang An-Shih, too, failed, partly because he had the imperious temperament of Gregory VII and Akhenaten and Caesar.

An outstanding example of success won by sympathetic understanding is the career of the Prophet Muhammad. He was so well attuned to the spirit of his time that he felt in himself, and perceived in his neighbours, the contemporary yearning of the Arab people for a monotheistic religion revealed in a scripture in the Arabic language. At the same time, Muhammad recognized that his mission was formidably difficult. The crux, for him, was the conversion of the Quraysh, the community that was in possession of Muhammad's native city-State, Mecca. Muhammad had two reasons for wanting to convert the Quraysh. They were the Prophet's own kinsmen, and they were an exceptionally sophisticated urban commercial community in an Arabian world of pastoral nomads. But the Quraysh believed that Muhammad's preaching of monotheism was a threat to their economic prosperity. The Quraysh made their living by trade; their trade depended on the prestige of the local sanctuary at Mecca, the Ka'bah; and the Ka'bah was dedicated, not to a single unique god, but to a pantheon.

Muhammad's sympathetic understanding was the quality that won for him his two decisive triumphs. When he was compelled to migrate, with a party of Meccan converts, from Mecca to Medina, he succeeded in coaxing the native Medinese and the Meccan *émigrés* to work together as a single community; and then, when he eventually compelled the anti-Islamic Meccans to capitulate and to embrace Islam, he offered them extraordinarily generous terms instead of penalizing them, now that they were at his mercy, for their obstinate resistance. By these two acts of sympathetic understanding, Muhammad ensured that Islam should survive him. By the time of his death, his former opponents at Mecca had come to realize that Islam, so far from ruining Mecca, was going to make Mecca's fortune. Consequently the Quraysh—including the majority of them that had been converted under duress at the eleventh hour—now threw themselves into the task of propagating Islam and extending the territory of the Islamic state. The rapid expansion of the Islamic state after Muhammad's death was due to the Quraysh's ability and experience, but it was Muhammad who had achieved the *tour de force* of enlisting the Quraysh for the Islamic state's service.

Trade requires negotiation; it is not possible to trade by force. Two of the principal traders among the Quraysh in Muhammad's generation were Abu Sufyān and his wife Hind, and their son Mu'āwīyah became Muhammad's fifth successor in the government of the Islamic state. Mu'āwīyah possessed the gift of sympathetic understanding that had been displayed by Muhammad and by Augustus and by Han Liu P'ang. By exercising this gift, Mu'āwīyah won away from his rival, the fourth successor of Muhammad, 'Alī, the support of a decisive majority of the Arab tribes who had been mobilized by Islam to serve as the Islamic state's army. Mu'āwīyah had the tact and the patience that were needed for reconciling the irascible and turbulent tribesmen to himself and to each other. He managed the Arabs as adroitly as the East Roman Emperor Aléxios I managed the Western Christian barons who passed through Constantinople, en route for Jerusalem, on the First Crusade.

Abraham Lincoln was a man of principle. But Lincoln was also sensitive and adroit. His objective was the abolition of slavery in the states in which slavery was still legal. He realized that the United States would not be able to preserve its political unity if slavery continued to be legal in some of the states after having been made illegal in the rest. He also realized that, in the states in which slavery was now illegal, public opinion was strongly in favour of preserving the Union, even at the cost of a civil war, but that it was not unanimous over the issue of the abolition of slavery, just because this issue was politically disruptive. Therefore, when the civil war broke out, Lincoln appealed for support for the preservation of the Union; he did not move for the abolition of slavery till he felt sure that, in the states that were fighting to preserve the Union, abolition had ceased to be a contentious question. Lincoln's perceptiveness and adroitness enabled him to secure both the preservation of the Union and the abolition of slavery.

Lastly, we have to consider the cases of leaders who did not wish to play this role. Saint Francis of Assisi rejoiced when his first adherent, Bernard of Quintavalle, joined him in espousing the Lady Poverty. But St Francis' objective was not the establishment of an institution; it was the imitation of Christ. When he found himself at the head of a numerous fraternity, he abdicated from his headship of the Franciscan Order and insisted on becoming a private soldier instead of continuing to be the commanding officer. Jesus, who was Francis' example, accepted the Jewish people's salutation of him as the Messiah, but he refused to be a belligerent Messiah, though, till then, this was the only kind of Messiah whom the Jewish people had imagined and desired. Did Jesus and Francis reject their opportunity? They did not accept it for themselves, but their self-denial inspired their followers to pursue the reluctant founder's objective on the founder's behalf. Much was retrieved, but something lost. Christians have not been Christs; Franciscans have not been Francises. Yet a precious aura of Christ's and Francis' spirit breathes in the institutions that bear their names. The opportunity has not been lost completely.

We may conclude that greatness and opportunity are always in need of each other; that their conjunction brings them both to fruition; and that, if they fail to meet, they will each remain barren.

THE FUTUROLOGIST

Johannes Witt-Hansen

The perception of the difference between 'before' and 'after' is one of the most pervasive experiences of daily life. The distinction between reversible and irreversible phenomena, introduced in physics by Carnot, Clausius and Boltzmann, uncovers new aspects of this experience, and discloses, as it were, its objective basis. As it turns out, the irretrievable losses of available energy, the trends towards increasing entropy or disorder in the universe where we live, make possible the distinction between an objective 'before' and an objective 'after' on both sides of an objective 'present'. Consequently, it would seem that this irreversibility is responsible for time slipping away irretrievably. No wonder that the irreversibility of definite physical events is accepted by physicists and philosophers at large as a crucial testimony to the conception that the modalities 'past time', 'present time', 'future time', including a definite irreversible direction, are fundamental aspects of objective time. In *The Nature of the Physical World* Eddington informs us that he shall use the phrase 'time's arrow' to express this one-way property of time which has no analogue in space. Adding that the great thing about time is that it goes on, he points out that the reversal of the arrow would render the external world nonsensical.[1]

The classification of natural phenomena in two classes, reversible and irreversible phenomena, is ignored or only vaguely foreshadowed by poets, philosophers and scientists in Antiquity. In Aristotle the reversal of 'time's arrow' is, in a certain sense, a universal law of 'things', and the idea of irreversibility of change in nature and society hardly even contemplated. The point is excellently made by Werner Jäger, who pictures this aspect of Aristotle's world-view in the following words:

'The things that change imitate those that are imperishable'. The coming-to-be and passing-away of earthly things is just as much a stationary revolution as the motion of the stars. In spite of its uninterrupted change nature has no history according to Aristotle, for organic becoming is held fast by the constancy of forms in a rhythm that remains eternally the same. Similarly the human world of state and society and mind appear to him not as caught in the incalculable mobility of irrecapturable historical destiny, whether we consider personal life or that of nations and cultures, but as founded fast in the unalterable permanence of forms that, while they change within certain limits, remain identical in essence and purpose. This feeling about life is symbolized by the Great Year, at the close of which all the stars have returned to their original position and begin their course anew.

Since, in Aristotle, time is the number of motion with respect to 'before' and 'after', it must consequently start from zero over and over again.[2]

In the same way the cultures of the earth wax and wane ... as determined by great natural catastrophes, which in turn are causally connected with the regular changes of the heavens. That which Aristotle at this instant newly discovers has been discerned a thousand times before, will be lost again, and one day discerned afresh.

As André Mercier[3] points out, Virgil anticipates unknowingly the second law of thermodynamics. In his poem on agriculture, *Georgics*, we find the passage: 'But time is on the move still, time that will not return, while we go cruising around this subject whose lore delights us' (*Sed fugit interea, fugit irreparabile tempus singula dum capti circumvectamur amore*).[4] Neither in the poet-scientist, Lucretius, nor in Archimedes is there any trace or presentiment of this law. Nevertheless, it is refreshing to learn from Lucretius' *De natura rerum* that 'time ... exists not of itself, but only from things is derived the sense of what has been done in the past, then what thing is present with us, further what is to follow after. Nor may we admit that anyone has a sense of time by itself, separated from the movement of things and quiet calm.'[5] Saint Augustine, the first futurologist of renown in Antiquity, advances in his theological language the view that 'there is no time before a universe exists' (*tempus nullum est ante mundum*), thereby supporting the view that time is dependent on the world of things. Although Saint Augustine expressly repudiates the view that time is identical with the motion of bodies, there is still a vagueness or ambiguity in his use of the term 'future' or 'future time'. Mostly the term 'future time' is understood as 'future calendar time'. In some cases, however, where he deals with society and history, the term refers to events coming-to-be (*omnia futura*). In his mentioning of God's foreknowledge of the future (*praescientia futurorum*), he has undoubtedly events in mind. This ambiguity, which is also deeply rooted in the vocabulary of modern futurologists, is so obvious that it rarely causes misunderstandings.

In his protest against the Hellenic cyclic conception of history and time, Saint Augustine advances a conception of history where social events have a definite direction, and time consequently is unidirectional and irreversible. A declaration to this effect is advanced in *Civitas Dei* in a famous passage:[6]

Philosophers of this world have thought that they could not or should not resolve this dispute [over the beginning of temporal events] in any other way than by introducing periodic cycles, in which, according to their contention, there has been an everlasting renewal and repetition of the same events in nature. ... But heaven forbid that our true faith should allow us to believe that these words of Solomon denoted those cycles in which, as those others think, the same measures of time and the same events in time are repeated in circular fashion: on the basis of this cyclic theory, it is argued, for example, that, just as in a certain age the philosopher Plato taught his students in the city of Athens and in the school called the Academy, so during countless past ages, at very prolonged yet definite intervals, the same Plato, the same city, and the same school with the same students had existed again and again, and during countless ages to come will exist again and again. Heaven forbid, I repeat, that we should believe that. For Christ died once for our sins, but 'rising from the dead he dies no more, and death shall no longer have dominion over him'.

Saint Augustine's conception of a unidirectional and irreversible course of history and time is rooted in the idea of a state of eternal happiness, guaranteed those who avoid the circuitous routes, whatever they are, and who follow the 'straight path of sound doctrine'. Such a conception implies that, even if there is in God's mind a definite pattern of causation, it does not follow that nothing is left to the free choice of the will. Translated into modern jargon, it would seem to mean that the choice of definite social initial states, which bring definite social laws into action, is left to those, namely the Christians, who are aware of the pattern. Since the future of human beings in the last analysis is a matter of choice, no course of events is inevitable. Saint Augustine protests vigorously against the view that the happiness held out in prospect to those who are redeemed should come to an end. There is no presentiment of a final state of decay, death and general disorder, no end of time. Obviously the course of historical events in Augustine is not only not in line with the cyclic movement in Aristotle or the Stoics. It also runs counter to the direction of natural phenomena towards decline and annihilation. And so does historical time. 'Time's arrow' is here an arrow of progress, directed towards a state of increasing order and a higher 'quality of life'. So far Saint Augustine anticipates modern futurology, or rather, some of its representatives.

Since modern research of futures is a global enterprise which has to be carried out conjointly with fellow scholars all over the world, it involves confrontations with philosophical views lurking in the background of social research in countries with different cultural patterns or political systems, and on different levels of industrial or economic development. Whereas there may be verbal consensus concerning the ultimate goals of humanity, the visions of the future or the futures of man, and the methods used for the research of futures are remarkably varied and manifold. Already before the modern trends in futurology were called into life, there existed in Western Europe a more or less philosophical interpretation of the future as a special existential dimension, extending from speculations of philosophers of history, like Kant, Hegel and Spengler, to the anticipations of utopians, sociologists, socialists and representatives of political science, like Saint-Simon, Fourier, Comte, Marx and Max Weber. In 1959 Teilhard de Chardin presented his eschatological view of the future in *L'Avenir de l'Homme*. In the same year Ernst Bloch published his *Das Prinzip Hoffnung*. Bertrand Russell, at an age of 89, envisaged the nuclear threat in *Has Man a Future?* In *The John Danz Lectures*,[7] 1964, Fred Hoyle broaches a subject of pressing importance at the present time. Almost unnoticed by futurologists, he describes the fatal race between our attempts to restrain our consumptive waste of available energy, and our attempt to acquire the information required for escaping our present global predicament. This race seems to be the core of the problem of 'future time', as far as human society and history are concerned. It leaves us at a crossroads where we have to choose between speeding up the collection of information securing, for a reasonable span of time, the trend towards increasing order within the bio- and sociosphere, and a very short 'golden age' of wasteful consumption of energy in some wealthy countries of the world, accelerating the flight of 'time's arrow' towards general disorder and disaster.

The point is that a choice between these alternatives is so hard to make because, for the present, no concerted decision-making on a world-wide scale is possible. Technically it is still feasible to reverse the process of depletion of resources, the consumptive waste of energy, and the growing disorganization of the biosphere. But politically it is next to impossible to do so. Since the problem is envisaged from several, and very different, points of view, there is no supreme court endowed with insight to demolish the myth of unlimited economic growth and prosperity, and equipped with political power to enforce political decisions in favour of the biosphere. In affluent societies it is tacitly assumed and accepted that such drawbacks or inconveniences as pollution, shortage of energy, warfare and racial unrest can be eliminated by concomitant preservation of an economic status quo. The idea of reversal of the process of wild economic growth and consumption of energy is obviously a threat to the gospel of welfare. Even representatives of definite trends in futurology—Arne Naess, Oslo, refers to them as The Shallow Ecology Movement[8]—are hardly aware that their assumptions are based on mutually exclusive demands, and that the basis of their doctrine of 'future time' is self-contradictory. Representatives of such technocratic-neoconservative trends in futurology are found among the initiators of research of futures in the United States (Hermann Kahn, Anthony Wiener, O. Helmer). However, much has changed lately, and for the present it is even good manners to denounce today's 'golden age' as tomorrow's nightmare. A new global existential philosophy is in *statu nascendi.* Here the worn-out *homo economicus* is gradually replaced by a human being worthy of bearing the predicates *sapiens* and *humanus.* We are informed that centuries of 'future time' are ruthlessly excised from the curriculum vitae of humanity to the extent in which resources are depleted and energy wasted. The basic facts of thermodynamics, information theory, and the status of the biosphere indicate that the present prevailing waste of energy in the 'rich countries' is a forfeiture of big lumps of the 'future time' of man. As Fred Hoyle warned:[10]

We have, or soon will have, exhausted the necessary physical prerequisites so far as this planet is concerned. With coal gone, oil gone, high-grade metallic ores gone, no species however competent can make the long climb from primitive conditions to high-level technology. This is a one-shot affair. If we fail, this planetary system fails so far as intelligence is concerned. The same will be true of other planetary systems. On each of them there will be one chance, and one chance only.

Let us look at our problem from another point of view, which takes us back to the history of social philosophy. Undoubtedly Marx was the most outstanding futurologist of the nineteenth century. Lenin, his spokesman in our century, linked Marxian ideas of socialism with the theory and practice of social planning. As Igor Bestuzhev-Lada[11] points out, Lenin emphasizes that the problems raised in historical materialism were concerned not only with the explanation of the past, but with boldly forecasting the future and taking practical action to bring it to pass. Indeed, the materialist conception of society and history is rather more concerned with the present and the future than with the past. In *Capital* it is announced that the ultimate aim of this work is to lay bare the economic law of modern (capitalist) society. Furthermore it is its aim to furnish a sort of forecast of the possibilities of transforming capitalism into a society rescued from the drawbacks of the anarchy of commodity production, capitalist style. Since historical materialism as interpreted by Lenin is the core of Soviet futurology, it is important for the understanding of its attitude to the present global situation to learn what this interpretation involves concerning the future of mankind. According to a wide-spread view, Marx established '"laws of development" ... which carry us irresistibly in a certain direction into the future. They are the basis of uncontrolled *prophesies*, as opposed to conditional scientific *predictions*'.[12] Popper even mentions 'the negation of the negation' as a 'dialectic law of motion' which 'furnishes the basis of Marx's prophesy of the impending end of capitalism'.[13] Although these views are wrongly attributed to Marx, they gain a certain support from a remark by Lenin in an article on Marx, where he says that 'it is evident that Marx deduces the inevitability of the transformation of capitalist society into socialist society wholly and exclusively from the economic law of motion of contemporary society'.[14] Referring to the development of capitalism since Marx's death, he says that this 'forms the chief material foundation for the inevitable coming of Socialism'.[15] If Lenin's remarks are taken at their face value, then his conception of the relation between social laws and forecasting or prediction differs fundamentally from that of Marx. The problem is: what interpretation of Marx is the basis of the 'ideological conflict between the world's socio-economic systems',[16] and how does it interfere with the Soviet futurologist's view of the future?

The problem is not an easy one. Marxist-Leninist ideology is mostly defined in terms that are not univocal, like 'scientific communism', 'the theoretical principles of dialectical and historical materialism', and so forth. Since the solution of the problem of ideological co-existence seems to be a necessary condition for a concerted decision-making on a world-wide scale, an elucidation of the problem of ideological conflict is of utmost concern to everybody engaged in forecasting, projecting, planning and designing the 'future time' of man. The strife of the parties centres around two opposing tenets: the 'everlasting existence' of capitalism, and the 'inevitability' of socialism or communism. As an argument for the communist tenet the following passage from the *Manifesto of the Communist Party* is often quoted: 'What the bourgeoisie ... produces, above all, are its own grave-diggers. Its fall and the victory of the proletariat are equally inevitable'.[17] Of course, these categorical statements are not predictions of future social events, but political slogans expressing the hopes which Marx and Engels certainly cherished. For the purpose of prediction and management within the capitalist society Marx established the laws of motion presented in *Capital*, whose role in the forecasting business is fundamentally different from the part played by political slogans.

Since the problem of the relation between Marxist conceptions of the future and Western conceptions of the subject has been placed in the forefront through an article by Pavel Apostol, Executive Director, Third World Future Research Conference, Bucharest, September 1972,[18] and a critical answer by the German futurologist, Ossip K. Flechtheim,[19] it seems worth while to try to put aside all political bias by which Marx's conceptions of social law, determinism, and the future have been stifled. Such an attempt might begin with the denunciation of three myths, closely interrelated: the first one is presented as the conception that the '15 sentences' from 'Preface' to *A Contribution to the Critique of Political Economy*[20] is a theory containing laws valid for all social formations. However, since these '15 sentences' are not connected by deductive links and do not include a proof structure, they do not constitute a theory in a strict sense. They are at most plausible assumptions, being part of the method of historical materialism. Consequently they cannot serve as bases for forecasts or predictions of social phenomena. Laws with such properties are, on the other hand, discovered in *Capital*, where at least two social theories, established according to the method of historical materialism, are presented: one which is very brief, namely the theory of an (idealized) social formation of simple commodity production, and a more extensive theory of the capitalist social formation, 'in its ideal average, as it were'.[21] Another interesting myth is summed up in the assertion that in Marx social development takes place according to a dialectical law expressed in the triad thesis-antithesis-synthesis. However, nowhere is it possible to discover in Marx's works the thesis-antithesis-synthesis paradigm as an element of his dialectic doctrine. No doubt the scheme was used by the Neo-Hegelians, and we know for certain that Proudhon used it in his *Philosophie de la Misère*. It is precisely in the *Misère de la Philosophie* that Marx derides Proudhon for his indiscriminate use of Hegelian and Neo-Hegelian slogans, and it is here we have to look for Marx's final relinquishment of such slogans. What matters to Marx is the Hegelian conception that any philosophical or scientific category represents or corresponds to a stage in the history of philosophy, science or political institutions. Since in Marx's view only the material social relations satisfy the criterion of recurrence, one must concentrate on having the economic categories correspond to definite stages of development of the material relations of production. This is the main point in Marx's conception of dialectics, set forth in a letter to J. B. Schweitzer, on 24 January 1865, a letter which deals with Proudhon's conception of the subject and his own view of the matter as presented in the *Misère de la Philosophie*:[22]

There I showed, among other things, how little he [Proudhon] had penetrated into the secret of scientific dialectics; how, on the other hand, he shares the illusions of speculative philosophy, for instead of conceiving the *economic categories as theoretical expressions of historical relations of production, corresponding to a particular stage of development of material production*, he garbles them into pre-existing, *eternal ideas*.

As it turns out, the economic categories or key-concepts which Marx has in view are value, money, corresponding to the relations of production of the social formation of simple commodity production; and surplus-value, capital, corresponding to the relations of production of the capitalist social formation. The dialectical process has, consequently, a double aspect: a logical and a historical one. A model pattern of such a process, illustrated by the transition from the social formation of simple commodity production to the capitalist social formation, is furnished in *Capital*. Here the logical aspect is presented in Part II, 'The Transformation of Money into Capital', whereas the historical aspect is described in Part VIII, 'The So-called Primitive Accumulation'.[23] Incidentally, it should be noted that the logical aspect of the dialectical process in Marx is essentially identical with the procedure known from mathematical and physical sciences as 'mathematical generalization'.[24]

Although the Hegelian jargon and the Hegelian terminology is pointless to Marx, he undoubtedly makes use of the terms and phrases of the dialectical vocabulary; and he does so in *Capital* in particular. The explanation of this odd fact is simply, as Marx informs the reader in the afterword to the second edition of *Capital*, that he wanted to call attention to the historical origin of the dialectical method in Hegel.[25] Of course it is out of the question that Marx considers an empty phrase like 'the negation of the negation' a basis for prophecy of the impending end of capitalism, as Popper puts it. This leads us to a third myth worth mentioning. It amounts to the assertion that social laws in Marx serve as a basis for prophecies. For its elucidation, and in order to throw some light on the Marxian conception of 'future time', it is necessary to discuss in some detail the problem of determinism in Marx and modern futurologists.

All futurologists agree that 'the future cannot be foreseen in the sense in which the various kinds of prophets understand the term'.[26] And nobody holds the view that the 'future time' of man is predictable in the way in which astronomical space-time events are predicted in classical mechanics. Since no convincing arguments have been advanced in favour of the view that a definite course of social events is inevitable, it seems reasonable to challenge the view that Marx deduces the inevitability of the transformation of capitalist society into socialist society from the economic laws of motion of modern society. However, the denial of this possibility imposes the obligation of identifying the part played by the laws of motion presented in *Capital*. It is known that Newton's method in physics was used to a great extent as a pattern of research by the Scottish moralists and economists, among others by Hume and Ferguson, and that Adam Smith in particular was heavily under its sway. This goes for Marx as well, who, impressed by the success of classical mathematics, attempted to make use of differential equations in his own variant of political economy. From several letters, and from the literary remains known as *Mathematical Manuscripts*,[27] it transpires that Marx planned a general mathematization of his laws of motion. He realized, however, that it could not be done for the time being. These laws were established in an analysis performed in an imaginary experiment with an idealized capitalist social formation, and on the empirical basis of the social statistics available at that time. It is often overlooked that the basic law of motion presented in *Capital*—a law concerning the motion of the rate of profit—is a hypothetical statement whose variables according to contemporary statistics have such values that the rate of profit has a tendency to fall. 'The hypothetical series drawn up at the beginning of this chapter expresses, therefore, the actual tendency of capitalist production.'[28] Although the stochastic nature of the law is not expressed mathematically, this property is illustrated verbally in terms of 'tendencies'. It would seem that these circumstances are sufficient reasons for rejecting the idea of the linearity of social development, attributed to Marx by authors from Kautsky and Plekhanov to T. W. Adorno. It also appears that the discussion of determinism in Marx has been on the wrong track insofar as the opposing parties in the strife, taking Laplacean determinism as a model pattern, seemed to ignore that the arbitrariness of the mechanical parameters determining the initial state of a mechanical system is as important a feature of the mechanical description as the deterministic character of the equations of motion.

As Danish physicist and philosopher Aage Petersen points out:[29]

> The logic of the mechanical description is often characterized in terms of Laplace's demon. Usually, the demon is pictured as a giant calculator who from knowledge of the state of the universe at a certain moment can calculate its state at any other time. However, in this version the simile is one-sided. The demon should rather be considered a chooser of mechanical parameters. Of course, this omnipotent experimenter should also have access to an ideal computer programmed to solve the mechanical equations of motion for any given choice of system and boundary conditions.

It is easy to agree with Aage Petersen that 'had Laplace stressed the role of the chooser, instead of that of the computer, subsequent philosophical discussions might have taken a different turn'.[30] This goes in particular for the strife about determinism in Marx. If it is accepted that the Marxian laws of motion have been constructed according to a modified Newtonian pattern, turning to account the logic of differential equations, it would seem that choice and arbitrariness have a place in Marxian sociology which has so far been ignored by politicians appealing to or opposing Marx. It is well known that differential equations, usual and partial, represent the most rigorous formulation of laws governing physical processes. This mathematical form enables us to make a clear-cut distinction between that which is grasped by a theoretical statement, and what is not. Due to their logical form they are absolutely silent concerning the future course of the physical systems to which they refer. Since in principle any arbitrary state of a system is possible, it is not until the initial state of a given system is chosen that the course which follows is determined. In Marx's case the laws of motion presented in *Capital* are intended to have similar logical properties. These laws are in themselves as silent concerning future social events as are the laws of motion of classical mechanics concerning the future history of a physical system. Only the choice of definite initial states, defined in terms of social statistics, makes possible the prediction of the probability of definite social events, for instance the probability of the fall or rise of the rate of profit. The analysis of the logical form of the laws of motion presented in *Capital* gives full support to the view that

Marx had cognizance of the logical status of social laws, and was aware of the role of choice of initial states in social forecasting. Since Marx did not formulate any law of transition from capitalist into socialist society, he did not envision the possibility of predicting the victory of communism in this or that country, or on a world-scale, on the basis of laws. As testified by Engels in the preface to the first German edition of *The Poverty of Philosophy*, Marx based his communist demands and the political hopes which he cherished upon historical studies and analogical-inductive reasoning:[31]

> According to the laws of bourgeois economics, the greatest part of the product does *not* belong to the workers who have produced it. If we now say: that is unjust, that ought not to be so, then that has nothing immediately to do with economics. We are merely saying that this economic fact is in contradiction to our sense of morality. Marx, therefore, never based his communist demands upon this, but upon the inevitable collapse of the capitalist mode of production which is daily taking place before our eyes to an ever greater degree; he says only that surplus-value consists of unpaid labour, which is a simple fact.

There is consequently no reason whatsoever to assume that Marx deduced the transformation of capitalist society into socialist society partly or exclusively from the laws of motion of contemporary society, or that these laws vindicate the belief in the inevitable coming of socialism. Nevertheless, since the thesis of inevitability of definite social events or phenomena still is part of socialist ideology, it remains part of socialist futurology as well.

The application of the laws of motion presented in *Capital* is a complicated affair because the problem of boundary conditions here is a very intricate one. This situation furnishes another reason for calling in question the possibility of deducing the inevitability of the coming of socialism from Marx's laws of motion. It should be recalled that Marx's analysis of the capitalist social formation starts with 'a very vague notion of a complex whole',[32] and stops by singling out for analysis proper definite social relations, i.e. 'relations of production', to which the criterion of recurrence can be applied. This goal is reached in a series of abstractive steps in a sort of imaginary experiment, where precisely the 'factors' lately brought into focus by futurologists[33] are 'thought away'. These factors are the population, its growth and mobility, the geographical environment, the status of the biosphere, etc. In Marx they are merely taken into account in terms of 'productive forces' and 'means of production', whereas their different rates of growth or decay in different cultures or areas of the planet are 'thought away'. Marx's view of the future is therefore limited within the boundaries which his method of abstraction or idealization has set. Consequently, the dependence of the growth of entropy in the biosphere on the production of consumption goods or the consumption of energy is not taken into account at all. Of course, a hundred years ago it was fully justified to shut one's eyes to the impact of the second law of thermodynamics on social science; and the increase of entropy or the growing disorder in our universe could safely be ignored. Today this is not so. The development of affluent societies these twenty-five years has produced biological and social phenomena surpassing our beliefs, expectations and forecasts to such an extent that the two basic tenets in the ideological strife between capitalism and socialism are severely shattered. It is now impossible to believe in 'eternal happiness' in everlasting affluent societies, feeding on the resources of developing countries; and face to face with the nuclear threat and a possible ecological catastrophe, it is hard to believe in the inevitability of any social system, whatever it may be. If it is true, as stressed by the Swiss ecologist Hans B. Barre at the Third World Future Research Conference, in Bucharest, in September 1972, that 'there is no difference in ideologies, when it comes to this problem: capitalism as well as socialism is still adhering to the religion of growth no matter how many sacrifices are required',[34] then there can hardly be any serious differences between capitalist and socialist futurologies concerning the objective in view. At the same conference Alexandro Nadal,[35] of Mexico, warned that the Western propagation of the ideology of consumption in the countries of the Third World gives rise to frustrations which, in view of their low standard of living, may produce explosive situations. Therefore, emphasis should be laid on the fulfilment of the basic requirements of food, clothing and dwelling, and on greater independence of the industrial countries. For the common future of mankind it is of decisive importance that it will be possible to give the population of the Third World a perspective of the future which indicates the way out of the present dilemma. A similar view is supported by Josué de Castro and Miguel A. Ozorio de Almeida, both of Brazil.[36] It would therefore seem that the basic ideological strife is not one between East and West, but a conflict between the rich countries and the rest of the world. This conflict is of course reflected in different conceptions of the basic problems of futurology, including the problem of time allotted to human beings on our planet. If this is so, we must acknowledge that futurology as a unitary discipline, taking into account all known conditions for the 'future time' of mankind, has just started to develop its own methods and its own philosophy. The different intuitive, explorative and projective methods,[37] exemplified in the Delphi-method, scenarios and simulations, are only just extensions of well-known methods used in sociology, economics, technological innovation and military strategies since the Second World War. Historical materialism as a method of research of futures has hardly transcended the limits set by Marx a hundred years ago. It may still be true, as pointed out by Bestuzhev-Lada in 1969,[38] that a specific methodology, i.e. a theory of forecasting in general and social forecasting in particular, is practically non-existent. A global existential philosophy, demanding a revolution in our way of thinking of human problems, is so far in *statu nascendi.* We are confronted with the problem of unravelling the logic of man's 'future time', including the logic of inevitability and possibility, necessity and contingency, determinateness and choice—a study still in its preparatory phase.

In Aristotle inevitability is attributed to the cyclic recurrence of inorganic, organic and social 'forms', and this goes for recurrent beginning of time at zero as well. In Virgil time slips away irretrievable. In Saint Augustine no course of social events is inevitable. Lenin deduces the inevitability of socialism from the laws of motion of capitalist society presented in *Capital.* Marx realizes vaguely that inevitability and necessity, whether causal or statistical, are logically distinct. His political hopes and demands, supported by Engels, take the form of slogans, according to which the collapse of capitalism and the victory of the proletariat appear as inevitable. In *Capital* he refers to future social events in terms of statistical necessity or 'tendencies', determined on the basis of laws of motion and political choice. Although the logic of our predicament is still unravelled, it is safe to say that, in our universe, the increase of disorder or entropy, whether macroscopic, mixed or thermodynamic, is inevitable. Whatever we do, we cannot stop the growth of this magnitude. So far, and since the reversal of 'time's arrow' is impossible, we are powerless in the face of nature's 'destructive forces'. Lately, however, we have become acquainted with the biosphere as a carrier of information and as an executor of a programme; moreover, the biological agent carrying out the ordering and adjustment activity attributed to Maxwell's demon has been identified. These ordering functions, which can be described in terms of mathematical statistics, are measurable in units of information and commensurable with those of thermodynamic entropy. In fact, we are inhabitants of a realm of growing order, negentropy of information, within which biological evolution and social development take place. This bio- and sociosphere, which also is a realm of contingency and necessity,[39] choice and determinateness, defies apparently the realm of inevitability or death, governed by the universal law of entropy. As pointed out by several authors[40] this would be a misrepresentation of the situation. The appropriation of information required to ensure order and stability, evolution and social development is possible only through consumption of equivalent amounts of available energy and subsequent production of entropy. As Léon Brillouin observes, the second law of thermodynamics is a decree of death, a decree which life defeats for the time being. Life acts on the fact that the decree of death is issued without settling the day of execution. And so does social life. This fact, to which futurologists have paid only little attention,[41] is most pertinent to the problem of man's common future[42] because it enables us, in principle, to compute the maximum and minimum of time allotted to man on our planet, and renders possible an estimate of the optimum and pessimum of the 'quality of life' within the time allotted.[43] If we see through the logic of 'future time' and accept a concerted decision-making on a world-wide scale, we may settle on postponement of the day of execution of the decree of death far beyond the year 2000. It becomes a matter of choice between a short-range and a long-range programme for mankind.

NOTES

1. A. S. Eddington, *The Nature of the Physical World*, p. 69, 68, Cambridge, Cambridge University Press, 1946.
2. W. Jäger, *Aristotle. Fundamentals of the History of his Development*, p. 389, London, Oxford University Press, 1962.
3. A. Mercier, *Fugit Irreparabile Tempus*, p. 5, 23, Berner Rektoratsreden. Bern, Verlag Paul Haupt.
4. Virgil. *The Eclogues, Georgics and Aeneid*, p. 98, London, Oxford University Press, 1966.
5. Lucretius. *De Natura Rerum*, book 1, 460–4, p. 34–5, London, William Heinemann (Loeb Classical Library); New York, N.Y., G. P. Putnam's Sons, 1924.
6. Saint Augustine, *The City of God Against the Pagans*, book XII, xiv, p. 58–61, 62–5, Cambridge, Mass., Harvard University Press; London, William Heinemann (Loeb Classical Library), 1966.
7. Fred Hoyle, *Of Men and Galaxies*, p. 49–72, London, Heinemann, 1965.
8. Arne Naess, 'The shallow and the deep long-range ecology movement', *Futuriblerne. Nordisk Tidsskrift for Fremtidsforskning*, Årgang 2, 1971/72, p. 172–4.
9. Editors of 'The Ecologist'. *A Blueprint for Survival*, Harmondsworth, Penguin Books Ltd, 1972; Donella H. Meadows, Dennis Meadows, Jørgen Randers, William W. Behrens, *The Limits to Growth*, London, Potomac Associates, Inc., 1972; Arne Naess, 'The Radical Message of Ecologists'. *SSRS Newsletter*, 231, April–May 1973, no. 3, p. 3–4.
10. Hoyle, op. cit., p. 64.
11. I. Bestuzhev-Lada, 'Forecasting—An Approach to the Problems of the Future', *International Social Science Journal*, Vol. XXI, No. 4, 1969, p. 527.
12. K. R. Popper, *The Poverty of Historicism*, p. 128, London, Routledge & Kegan Paul, 1961.
13. K. R. Popper, *Conjectures and Refutations. The Growth of Scientific Knowledge*, p. 333, London, Routledge & Kegan Paul, 1963.
14. V. I. Lenin, *Karl Marx*, p. 49–50, Moscow, Foreign Languages Publishing House, 1951.
15. ibid.
16. Bestuzhev-Lada, op. cit., p. 534.
17. K. Marx and F. Engels, *Selected Works*, Vol. I, p. 43, Moscow, Foreign Languages Publishing House, 1950.
18. P. Apostol, 'Marxism and the Structure of the Future', *Futures*, September 1972, p. 201–10.
19. Ossip K. Flechtheim, 'Eine Antwort auf die Zukunftsvorstellungen eines marxistischen Futurologen', *Analysen und Prognosen über die Welt von Morgen*, No. 26, March 1973, p. 19–21.
20. Marx and Engels, op. cit., p. 328–9.

21. K. Marx, *Capital. A Critique of Political Economy*, Vol. III, p. 831 (edited by Friedrich Engels), New York, N.Y., International Publishers, 1967.
22. K. Marx and F. Engels, *Selected Correspondence*, p. 172, London, Lawrence & Wishart, 1943.
23. K. Marx, *Capital. A Critique of Political Economy*, Vol. I, op. cit., p. 146–76, 713–65.
24. J. Witt-Hansen, 'The Impact of Bohr's Thought on Danish Philosophy', in: Raymond E. Olson and Anthony M. Paul (eds.), *Contemporary Philosophy in Scandinavia*, p. 491–508. Introduction by G. H. von Wright. Baltimore and London, The Johns Hopkins Press, 1972.
25. K. Marx, *Capital. A Critique of Political Economy*, Vol. I, op. cit., p. 19–20.
26. Bestuzhev-Lada, op. cit., p. 530.
27. K. Marx, *Matematičeskie Rukopisi*, Moscow, Izdatel'stvo 'Nauka', Glavnaja Redakcija Fiziko-matematičeskoj Literatury, 1968.
28. K. Marx, *Capital. A Critique of Political Economy*, Vol. III, op. cit., p. 50, 212.
29. A. Petersen, *Quantum Physics and the Philosophical Tradition*, p. 150–1, New York, N.Y., Belfer Graduate School of Science, Yeshiva University (dissertation), 1968.
30. ibid.
31. K. Marx and F. Engels, 'Preface', in *The Poverty of Philosophy. Answer to the 'Philosophy of Poverty' by M. Proudhon*, p. 9, Moscow, Progress Publishers, 1966.
32. K. Marx, *A Contribution to the Critique of Political Economy*, p. 205. London, Lawrence & Wishart, 1971.
33. Jay W. Forrester, *World Dynamics*, Cambridge, Mass., Wright-Allen Press, Inc., 1971.
34. Hans B. Barre, *Towards an International Equilibrium Year*, p. 3, Report to the Third World Future Research Conference, Bucharest, September 1972.
35. Gunther Heyder, 'Strategien für eine humane Welt. Bericht über die 3. Weltkonferenz für Zukunftsforschung in Bukarest', *Analysen und Prognosen über die Welt von Morgen*, November 1972, No. 24. p. 25–6.
36. Josué de Castro, 'Polution Problem No. 1—Under Development'. *The Unesco Courier*, January 1973, p. 20–3; Miguel A. Ozorio de Almeida. 'The Myth of Ecological Equilibrium', ibid., p. 25–8.
37. Ossip K. Flechtheim, *Futurologie. Der Kampf um die Zukunft*, p. 85–104, Frankfurt-am-Main, Fischer Taschenbuch Verlag GmbH and Köln, Verlag Wissenschaft und Politik, 1972.
38. I. Bestuzhev-Lada, 'Forecasting—An Approach to the Problem of the Future', *International Social Science Journal*, Vol. XXI, No. 4, 1969, p. 530.
39. Jacques Monod, *Le Hasard et la Necessité. Essais sur la Philosophie Naturelle de la Biologie Moderne*, Paris, Éditions du Seuil.
40. Léon Brillouin, *Science and Information Theory*, 2nd ed., New York, N.Y., Academic Press Inc., 1962; P. Chambadal, *Évolution et Applications du Concept d'Entropie*, Paris, Dunod, 1963; K. S. Trinčer, *Biologija i Informacija. Elementy Biologičeskoj Termodinamiki* [*Biology and Information. Elements of Biological Thermodynamics*], Moscow, Izdatel'stvo 'Nauka', 1965.
41. University of Sussex Science Policy Research Unit (United Kingdom) (with a response by Dennis Meadows *et al.*), 'The Limits to Growth Controversy. World Dynamics Models Described and Evaluated—Resources, Population, Agriculture, Capital, Pollution, Energy', *Futures*, Vol. 5, No. 1, February 1973.
42. Johan Galtung, 'Man's Common Future. Suggested Theme for the 3rd International Future Research Conference', *Futuriblerne. Nordisk Tidsskrift for Fremtidsforskning*, Årgang 2, 1971, 3, p. 87–9.
43. Michel Grenon, *Ce Monde Affamé d'Énergie* (Foreword by Sicco Mansholt), Paris, Laffont, 1973.

[A.45] SS.76/XXII 2/A